Marco Iacoboni is a neurologist and neuroscientist
originally from Italy. Currently he is at the David
Geffen School of Medicine, UCLA, where he is
director of the Transcranial Magnetic Stimulation
Laboratory at the Ahmanson-Lovelace Brain Mapping Center. His brain imaging studies have pio-
neered the investigation of the mirror neuron
system in humans. He lives in Los Angeles, California.

MIRRORING PEOPLE

THE SCIENCE OF EMPATHY AND HOW
WE CONNECT WITH OTHERS

MIRRORING PEOPLE

MIRRORING PEOPLE

MARCO IACOBONI

Picador

———

Farrar, Straus and Giroux

New York

www.picadorusa.com

Picador® is a U.S. registered trademark and is used by Farrar, Straus and Giroux under license from Pan Books Limited.

For information on Picador Reading Group Guides, please contact Picador.
E-mail: readinggroupguides@picadorusa.com

Figures of brain areas with mirror neurons in macaques and humans courtesy of Sinauer Associated, Inc.

Designed by Michelle McMillian

ISBN-13: 978-0-312-42838-9
ISBN-10: 0-312-42838-3

First published in the United States by Farrar, Straus and Giroux as *Mirroring People: The New Science of How We Connect with Others*

First Picador Edition: July 2009

10 9 8 7 6 5 4 3 2 1

To my wife, Mirella, my daughter, Caterina,
and my parents, Rita and Antonio

Contents

CONTENTS

CONTENTS

CONTENTS

Ten: Neuropolitics

Eleven: Existential Neuroscience and Society

MIRRORING PEOPLE

ONE

Monkey See, Monkey Do

NEURO THIS!

When we get right down to it, what do we human beings do all day long? We *read the world*, especially the people we encounter. My face in the mirror first thing in the morning doesn't look too good, but the face beside me in the mirror tells me that my lovely wife is off to a good start. One glance at my eleven-year-old daughter at the breakfast table tells me to tread carefully and sip my espresso in silence. When a colleague reaches for a wrench in the laboratory, I know he's going to work on the magnetic stimulation machine, and he's not going to throw his tool against the wall in anger. When another colleague walks in with a grin or a smirk on her face—the line can be fine indeed, the product of tiny differences in the way we set our face muscles—I automatically and almost instantaneously can discern which it is. We all make dozens—hundreds—of such distinctions every day. It is, quite literally, what we *do*.

Nor do we give any of this a second thought. It all seems so ordinary. However, it is actually extraordinary—and extraordinary that it feels ordinary! For centuries, philosophers scratched their heads over humans' ability to understand one another. Their befuddlement was reasonable: they had essentially no science to work with. For the past 150 years or so, psychologists, cognitive scientists, and neuroscientists have had some science to work with—and in the past fifty years, a lot of science—and for a long time they continued to scratch their heads. No one could begin to explain how it is that we know what others are doing, thinking, and feeling.

Now we can. We achieve our very subtle understanding of other people thanks to certain collections of special cells in the brain called mirror neurons. These are the tiny miracles that get us through the day. They are at the heart of how we navigate through our lives. They bind us with each other, mentally and emotionally.

Why do we give ourselves over to emotion during the carefully crafted, heartrending scenes in certain movies? Because mirror neurons in our brains re-create for us the distress we see on the screen. We have empathy for the fictional characters—we know how they're feeling—because we literally experience the same feelings ourselves. And when we watch the movie stars kiss on-screen? Some of the cells firing in our brain are the same ones that fire when we kiss our lovers. "Vicarious" is not a strong enough word to describe the effect of these mirror neurons. When we see someone else suffering or in pain, mirror neurons help us to read her or his facial expression and actually make us feel the suffering or the pain of the other person. These moments, I will argue, are the foundation of empathy and possibly of morality, a morality that is deeply rooted in our biology. Do you watch sports on television? If so, you must have noticed the many "reaction shots" in the stands: the fan frozen with anticipation, the fan ecstatic over the play. (This is especially true for baseball broadcasts, with all the downtime between pitches.) These shots are effective television because our mirror neurons make sure that by seeing these emotions, we *share* them. To see the athletes perform is to perform ourselves. Some of the same neurons that fire when we watch a player catch a ball also fire when we catch a ball ourselves. It is as if by watching, we are also playing the game. We understand the players' actions because we have a template in our brains for that action, a template based on our own movements. Since different actions share similar movement properties and activate similar muscles, we don't have to be skilled players to "mirror" the athletes in our brain. The mirror neurons of a non-tennis-playing fan will fire when watching a pro smash an overhead, because the non-tennis-playing fan has certainly made other kinds of overhead movements with his arm throughout his life; the equivalent neurons of a fan such as me, who also plays the game, will obviously be activated much more strongly. And if I'm watching Roger Federer, I bet my mirror neurons must be firing wildly, because I'm a big Federer fan.

Mirror neurons undoubtedly provide, for the first time in history, a plausible neurophysiological explanation for complex forms of social cognition and interaction. By helping us recognize the actions of other people, mirror neurons also help us to recognize and understand the deepest motives behind those actions, the intentions of other individuals. The empirical study of intention has always been considered almost impossible, because intentions were deemed too "mental" to be studied with empirical tools. How do we even know that other people have mental states similar to our own? Philosophers have mulled over this "problem of other minds" for centuries, with very little progress. Now they have some real science to work with. Research on mirror neurons gives them and everyone interested in how we understand one another some remarkable food for thought.

Consider the teacup experiment I dreamed up some years back, which I'll discuss in considerable detail later. The test subjects are shown three video clips involving the same simple action: a hand grasping a teacup. In one, there is no context for the action, just the hand and the cup. In another, the subjects see a messy table, complete with cookie crumbs and dirty napkins—the aftermath of a tea party, clearly. The third video shows a neatly organized tabletop, in apparent preparation for the tea party. In all three video clips, a hand reaches in to pick up the teacup. Nothing else happens, so the grasping action observed by the subjects in the experiment is exactly the same. The only difference is the context.

Do mirror neurons in the brains of our subjects note the difference in the contexts? Yes. When the subject is observing the grasping scene with no context at all, mirror neurons are the least active. They are more active when the subject is watching either of the scenes and *most active* when watching the neat scene. Why? Because drinking is a much more fundamental intention for us than is cleaning up. The teacup experiment is now well known in the field of neuroscience, but it is not an isolated result: solid empirical evidence suggests that our brains are capable of mirroring the deepest aspects of the minds of others—intention is definitely one such aspect—at the fine-grained level of a *single brain cell*. This is utterly remarkable. Equally remarkable is the effortlessness of this simulation. We do not have to draw complex inferences or run complicated algorithms. Instead, we use mirror neurons.

Looking at the issue from another perspective, labs around the world are accumulating evidence that social *deficits*, such as those associated with autism, may be due to a primary *dysfunction* of mirror neurons. I hypothesize that mirror neurons may also be very important in imitative violence induced by media violence, and we have preliminary evidence suggesting that mirror neurons are important in various forms of social identification, including "branding" and affiliation with a political party. Have you heard of neuroethics, neuromarketing, neuropolitics? You will in the years and decades to come, and research in these fields will be rooted, explicitly or otherwise, in the functions of mirror neurons.

This book tells the story of the serendipitous and groundbreaking discovery of this special class of brain cells, the remarkable advances in the field in just twenty years, and the extremely clever experiments now under way in several labs around the world. Quite simply, I believe this work will force us to rethink radically the deepest aspects of our social relations and our very selves. Some years ago, another researcher suggested that the discovery of mirror neurons promised to do for neuroscience what the discovery of DNA did for biology.[1] That's an extraordinarily bold statement, because essentially everything in biology comes back to DNA. Decades in the future, will everything in neuroscience be seen as coming back to mirror neurons?

BRAIN SURPRISES

For fifteen years I have lived in Los Angeles and worked in my laboratory at UCLA, but as my name suggests, this story should rightly begin in Italy, and I'm happy to report that it really does—specifically, in the small and beautiful city of Parma, famous for its fabulous food, particularly prosciutto di Parma and Parmesan cheese, and for its music. Now we can add neuroscience to the list of Parma's world-class exports; it was at the university here that a group of neurophysiologists, led by my friend Giacomo Rizzolatti, first identified mirror neurons.

Rizzolatti and his colleagues work with *Macaca nemestrina*, a species of monkey often used in neuroscience labs worldwide. These macaque monkeys are very docile animals, unlike their more famous relatives, rhesus monkeys, who are highly competitive alpha-male types (even the females). Research on monkeys in a lab such as Rizzolatti's is predicated on its inferential value for understanding the human brain, which is generally considered the most complex entity in the known universe, with good reason. The human brain contains about one hundred billion neurons, each of which can make contact with thousands, even tens of thousands, of other neurons. These contacts, or synapses, are the means by which neurons communicate with one another, and their number is staggering. The distinguishing brain feature in mammals is the neocortex, the most recently evolved of our brain structures. Now here's the key "inferential" point: the macaque brain is only about one-fourth as large as ours, and our neocortex is much larger than the macaque's neocortex, but neuroanatomists typically agree that the structures in the neocortex of macaques and humans correspond relatively well despite these differences.

In Parma, the Rizzolatti team's pertinent area of study was an area of the brain labeled F5, located in a large region called the premotor cortex—that part of the neocortex concerned with planning, selecting, and executing actions. Area F5 contains millions of neurons that specialize in "coding" for one specific motor behavior: actions of the hand, including grasping, holding, tearing, and, most fundamental of all, bringing objects—food—to the mouth. For every macaque, as for every primate, these actions are as basic and essential as they come. We *Homo sapiens* are grasping and manipulating objects from the moment we fumble for the snooze button on the alarm clock until we adjust our pillows at bedtime, eighteen hours later. All in all, we each perform hundreds, if not thousands, of grasping actions every day. In fact, this is precisely why the Rizzolatti team chose area F5 for the closest possible investigation. All neuroscientists want to understand the brain for understanding's sake, but we also have an eye on more practical goals, such as discoveries that may eventually drive new treatments for disease. The discovery of the neurophysiological mechanisms of motor control of the hand in the macaque could eventually help individual humans with brain damage recover at least some degree of hand function.

Through laborious experimentation, the Rizzolatti team had acquired an impressive understanding of the actions of these motor cells during various "grasping" exercises with the monkeys. (They are called motor cells because they are the first in the sequence that controls the muscles that move the body.) Then one day, about twenty years ago, the neurophysiologist Vittorio Gallese was moving around the lab during a lull in the day's experiment. A monkey was sitting quietly in the chair, waiting for her next assignment. Suddenly, just as Vittorio reached for something—he does not remember what—he heard a burst of activity from the computer that was connected to the electrodes that had been surgically implanted in the monkey's brain. To the inexperienced ear, this activity would have sounded like static; to the ear of an expert neuroscientist, it signaled a discharge from the pertinent cell in area F5. Vittorio immediately thought the reaction was strange. The monkey was just sitting quietly, not intending to grasp anything, yet this neuron affiliated with the grasping action had fired nevertheless.

Or so goes one story about the first recorded observation of a mirror neuron. Another involves one of Vittorio's colleagues, Leo Fogassi, who picked up a peanut and triggered an excited response in F5. Yet another credits Vittorio Gallese and some ice cream. There are others, all plausible, none confirmed. Years later, when the importance of mirror neurons was clearly understood, the Parma colleagues went back to their lab notes, hoping to put together a fairly accurate timeline of their earliest observations, but they simply couldn't do it. They found references in their lab notes to "complex visual responses" of the monkeys' motor cells in area F5. Such notes were unclear, because the scientists did not know what to make of their observations at the time. Neither they nor any neuroscientist in the world could have imagined that motor cells could fire merely at the *perception* of somebody else's actions, with no motor action involved at all. In light of both knowledge and theory at the time, this made absolutely no sense. Cells in the monkey brain that send signals to other cells that are anatomically connected to muscles have no business firing when the monkey is completely still, hands in lap, watching somebody else's actions. And yet they did.

In the end, it does not matter much that the "Eureka!" moment for mirror neurons stretched over a period of years. What matters is that the team did soon grapple with the odd goings-on in their laboratory. They had a hard time believing these phenomena themselves, but in time they also sensed that the discovery, if confirmed, could be potentially groundbreaking. They were right. Twenty years after that first recording in the laboratory, a cascade of well-controlled experiments with monkeys and later with humans (different kinds of experiments, for the most part; no needles inserted through skulls) have confirmed the remarkable phenomenon. The simple fact that a subset of the cells in our brains—the mirror neurons—fire when an individual kicks a soccer ball, sees a ball being kicked, hears a ball being kicked, and even just says or hears the word "kick" leads to amazing consequences and new understandings.

THE FAB FOUR

We now know that about 20 percent of the cells in area F5 of the macaque brain are mirror neurons; 80 percent are not. Given those odds, the group in Parma was bound to come across mirror neurons sooner or later. When they did, the background assumptions not only of their lab but of neuroscientists around the globe were put to the test. In the 1980s, neuroscientists were deeply invested in the paradigm that held that the various functions implemented by the brain—macaque or human—were confined to separate boxes. Under this paradigm, perception (seeing objects, hearing sounds, and so on) and action (reaching for a piece of food, grasping it, putting it in the mouth) are entirely separate and independent of one another. A third function, cognition, is somehow "in between" perception and action and allows us to plan and select our motor behavior, to attend to specific things that are relevant to us, to disregard extraneous matters, to remember names and events, and so on. These three broadly construed functions were typically assumed to be separate in the brain. The paradigm reflected the justified bias in science for the most parsimonious explanation of phenomena. To dissect a complex phenomenon into simpler elements is a good principle for investigation. It is still the dominant approach in neurophysiology and neuroscience, and in many specialized areas of research it works well. For instance, researchers have identified neurons that respond only to horizontal lines in the visual field, while others code for vertical lines.

Many brain cells do seem to be highly and narrowly specialized. The neuroscientist who assumes that neurons can be so easily categorized, however—with no crossing over between perception, action, and cognition—may miss entirely (or dismiss as a fluke) neuronal activity that codes with much more complexity, that reflects a brain that is dealing with the world in a much more "holistic" fashion than previously understood. Such was the case with mirror neurons. The Parma investigators, each and every one of whom was a superb scientist, were nevertheless unprepared for a motor neuron that was also a perception neuron. An old quip makes this point in general terms: "Progress in science proceeds one funeral at a time." That is rather morbid and also a great exaggeration, but we all know that it is hard to give up the old paradigm, to think outside the box, to change—and not just in science. Indeed, quite a few years were required for them (and, by then, other investigators around the world) to figure out the "complex visual responses" recorded in the lab. Initially, scientists were not mentally ready to challenge the assumptions inherited from generations of researchers; those assumptions had guided a lot of productive research. Moreover, no findings prior to that moment had contradicted the assumptions.[2]

Now they did—and in more ways than one. During the years of early work with mirror neurons, the Rizzolatti team

was also uncovering a set of cells in area F5 with another feature for which they could not account. These were cells that
fired during grasping behavior and also at the very sight of
graspable objects. They were later called canonical neurons, a
bit ironically. Both of these patterns of neural activity contra-
dict the old idea that action and perception are completely
independent processes confined to their separate boxes in the
brain. In the real world, as it turns out, neither the monkey
nor the human can observe someone else picking up an apple
without also invoking in the brain the motor plans necessary
to snatch that apple themselves (mirror neuron activation).
Likewise, neither the monkey nor the human can even look
at an apple without also invoking the motor plans necessary
to grab it (canonical neuron activation). In short, the grasping actions and motor plans necessary to obtain and eat a
piece of fruit are inherently linked to our very *understanding* of
the fruit. The firing pattern of both mirror and canonical neurons in area F5 shows clearly that perception and action are
not separated in the brain. They are simply two sides of the
same coin, inextricably linked to each other.

Some of the earliest macaque experiments in Parma—back
in the 1980s, years before the puzzling episodes that turned
out to mark the discovery of mirror neurons—supported these
same conclusions about the tight link between perception
and action. At that time, the team conducted a set of experiments focused not on area F5 in the motor cortex, but on
the adjacent area F4. In area F5, as we have seen, the cells
most readily fire while the monkey performs actions with
the hands. Neurons in area F5 also fire when the monkey

makes mouth actions such as biting, as well as facial communicative gestures such as lip smacking, which has a positive social meaning in primates.[3] Indeed, some neurons in F5 fire for hand actions *and* for mouth actions. The firing pattern of these neurons is yet another feature that contradicts those models of the brain made of separate boxes, with one box for the hand and a separate one for the mouth. (This is how an engineer, I guess, would probably build a brain.) Neurons that code for both hand and mouth actions, however, make perfect sense to holistic interpretations of brain functions, in which motor cells are concerned with the *goal* of an action. Indeed, the hand brings the food to the mouth. In area F4, the cells fire mostly while the monkey moves the arm, the neck, and the face. That was the thinking, and those were the results of the experiments before the discovery that the cells also fire in response to sensory stimulation alone, with no movement from the monkey. Moreover, they respond to stimulation only from real objects. Simple lights or shapes projected on a screen do not trigger any discharge from these cells. Also, they fire only when the objects in question are quite close to the monkey's body, and they fire more strongly when the objects are rapidly approaching. Another peculiar feature of these cells: they respond to a simple touching of the face, neck, or arm of the monkey. Conclusion: the visual receptive field (that part of the surrounding space in which visual stimuli trigger the firing of the cell) and the tactile receptive field (that part of the body that, when touched, triggers the firing of the cell) are related in these neurons in area F4. Their amazing responses suggest that they are creating a map of the space surrounding the body, what we call a peripersonal space map. And they also trigger the monkey's movement of the arm, say, in that space. Two very different functions are manifested in one group of cells. These physiological properties suggest that the space map surrounding the body is a map of *potential actions* performed by the body.[4]

As it happens, the new paradigm inaugurated by the discovery of these F4 and F5 neurons—including mirror neurons, of course—was foreseen in a way by Maurice Merleau-Ponty, a French philosopher working in the early twentieth century. Merleau-Ponty was a member of a school of philosophy known in the decades around 1900 as phenomenology. Other members were Franz Brentano, Edmund Husserl, and the great Martin Heidegger. They criticized the classical philosophical approach as being seduced by the holy grail of discovering the very essence of phenomena and thus getting bogged down in musing about abstractions (the Platonic tradition), and they proposed instead to "go back to the things themselves" (the Aristotelian instinct, in effect). The phenomenologists proposed to pay close attention to the objects and phenomena of the world and to our own inner experience of these objects and phenomena. In the lab in Parma, Rizzolatti and his colleagues were very traditional in the techniques they used to study the cells in areas F4 and F5 in the frontal cortex of their macaque monkeys, but over time, they were able to overcome the traditional framework for interpreting their results: separate compartments for motor, perception, and cognition cells. They were able to wipe away the ruling paradigm and hypotheses. They didn't waste years trying to extract complex and abstract computational rules to explain the apparently bizarre observations that were piling up. Instead, they were able to employ a fresh, open-minded approach to the research, which I call neurophysiologic phenomenology. Their new attitude was the only means of realizing that perception and action are a unified process in the brain.

The head "philosopher" in Parma was the bearded and dark-eyed neurophysiologist Vittorio Gallese. Gallese was the one digging into Merleau-Ponty's work, finding the appropriate analogies between philosophy and neuroscience, explaining the group's discoveries in less scientific and more philosophical terms. Gallese was also more willing to speculate about the most profound implications of mirror neurons. Indeed, his presentation at a meeting titled "Toward a Science of Consciousness," in Tucson, Arizona, in 1998, was the catalyst for making mirror neurons well known in the scientific world for the first time. At that meeting Gallese met coincidentally Alvin Goldman, a philosopher interested in the problem of other minds. Goldman is a paladin of the simulation theory, which holds that in order to understand what another person feels when, say, she is in love, we must pretend to be in love ourselves. He immediately caught the implications of this new mirror neuron research for his own thinking, and he and Gallese worked together on a paper that proposed for the first time that mirror neurons may be the neural correlate of the simulation process necessary to understand other minds.[5]

Gallese's passion for philosophy and science is surpassed

only by his love for the opera, which is not at all unusual in Parma. He is one of the twenty-seven members of the exclusive Club dei 27 (www.clubdei27.com), wherein each member personifies one of Giuseppe Verdi's twenty-seven operas. I do not use the term "exclusive" lightly. There will be no more operas from Verdi, may he rest in peace, so it will never be Club dei 28. The only way one can become a member is for another member to pass the torch (highly unlikely) or pass away. Gallese personifies a lesser-known opera of the Maestro, I Lombardi alla prima crociata, but of course he had no choice in this. He grabbed the only available opening! A highlight of Gallese's third career (neuroscience and philosophy are the first two) was the night that the Club dei 27 bestowed a medal on the peerless Spanish tenor Plácido Domingo. Gallese joined his twenty-six confreres in singing for the listening pleasure of one of the greatest of all Verdi interpreters.

Is that preceding paragraph a digression? I don't think so. With the rarest of exceptions, great science is about the combined, dogged legwork of at least several if not many individuals. It's all about teamwork. And what makes a great team of any sort? Nobody really knows, but when it happens, anyone can clearly see the results. In the lab in Parma directed by Giacomo Rizzolatti, a collection of neuroscientists contributed to the magic in many different ways. Vittorio Gallese's interest in philosophy and phenomenology was not incidental at all; in fact, it was probably critical. His philosophical tendency and his passion for the opera are markers of a personality with broad interests and an ability and willingness to think outside the box. In my experience, the best scientists are *interesting people*.

Joining Gallese and the lab director, Rizzolatti, as key members of the team were Luciano Fadiga and Leo Fogassi. These four neuroscientists are all quite different from one another in personality and intellectual inclinations. Maybe this is one of the reasons things worked out so well. In any event, each made a unique contribution to the collective work, as would be the case in any world-class scientific enterprise. Fadiga, tall and slender, has a knack for developing new tools for the laboratory and possesses the social skills necessary for management and fund-raising. Modern science requires all three: technological innovation, management skills, and lots of money. (With machines that can easily cost hundreds of thousands of dollars or as much as two or three million, basic research in neuroscience is particularly expensive.) Typically, the scientists who are great with technical matters in the lab are not so good at the "people" end of the job. Fadiga is one of the exceptions to this rule. He was the team member who first applied the relatively new technique of transcranial magnetic stimulation (TMS) to the study of the mirror neuron system in humans (a subject for later discussion). He recently moved to the University of Ferrara, where his new lab is already an efficient and productive machine. No surprise there.

In contrast with Fadiga, Leo Fogassi is by far the least outspoken individual of the four Parma neuroscientists. In the years just after the discovery of mirror cells in the early 1990s, Fogassi was definitely less involved in communicating the experimental findings to the scientific community. Communication is a fundamental aspect of science, of course, but it is just not Fogassi's forte. He's a great lab man, probably the one scientist who has directly performed or supervised the largest number of single-cell recordings in the mirror neuron system in the world. In the last few years he has taken the lead in a variety of important projects, most important of which is the series of experiments on the role of mirror neurons in understanding the intentions of others. I'll discuss this vital work shortly.

This brings us to the leader of the group, Giacomo Rizzolatti, who should be considered as nothing less than a Renaissance man. In modern science, specialization is the order of the day, and then specialization within specialization. Most scientists focus on a single research issue, using just one modality of investigation. Rizzolatti's research ranges far and wide, including visual neurophysiology in cats, behavioral neurology in brain-damaged patients, experimental psychology in healthy volunteers, anatomical and neurophysiological studies in primates, brain imaging in humans, and—in addition to all this—computational neuroscience! Rizzolatti's ability to connect all these different lines of research in an integrated and coherent view of human brain function is almost uncanny and definitely unique in modern neuroscience. Above all, his intuitions about how the brain works are incomparable. (Maybe this talent for deep insight is why his somewhat ruffled white hair always reminds me of Albert Einstein.) The early work in Parma, which led to the discovery of mirror neurons, originated from Rizzolatti's intuitions about the role of premotor areas in creating "space maps" surrounding the body. He called this theory the premotor theory of attention. Several years ago, by simply looking at the pattern of reaction-time data in healthy volunteers during a visuospatial task (certainly not the most self-explanatory piece of information with regard to brain function), Rizzolatti proposed a model of visuospatial attention—that is, how we pay attention to some object or movement on our left side and not on our right side—that was confirmed by brain imaging techniques many years later.[6]

Rizzolatti, Gallese, Fogassi, and Fadiga were the Fab Four; and together they changed everything. The discovery of mirror neurons and the development of their possible implications was primarily due to the collaborative chemistry of these four neuroscientists. In the years to come, even the educated layperson's understanding of how we humans really see the world, and how we function as social animals within it, will never be the same.

MIRRORS IN THE BRAIN

Isn't the devil in the details? In neuroscience, at least, this always seems to be the case, and it is definitely so with mirror neurons. It is the slight variations of the experimental setups in labs around the world that have revealed the subtlety of responses from these neurons, which in turn have opened the doors to our understanding. On the other hand, there was nothing unique about the investigative tools used in Parma.

Using the classic methodology of single-cell neurophysiology, Rizzolatti and his colleagues implanted electrodes in area F5 of the macaque subjects and recorded any electrical changes—"action potentials"—on the surface of individual neurons as the monkeys performed certain tasks in exchange for rewards of food. The electrical activity in the brain is what tells us that a particular neuron has been activated at a particular time. It has "fired," as we say, and it has done so in order to code either a sensory event (seeing some object or action), a motor act (grasping the apple), or a cognitive process (the memory of grasping the apple). (Under the old "separate boxes" paradigm, as we have seen, any given cell would code for one and only one of these three activities. Mirror neurons code for two of them, breaking down the barrier between perception and action.) These electrical discharges are also the way brain cells send signals to one another. Even cells that are far from each other in the brain can communicate through action potentials, as long as they are physically connected with axons, which are long extensions of the cell that serve as extension cords of a sort.

These classical experiments give us access to brain activity at its most exquisite, fine-grained level—the single cell—and provide exquisite spatial and temporal "resolution." We are working not only with the single cell, but instant by instant. This research provides incredibly important information. From our understanding of the brain mechanisms of our evolutionary predecessors we can infer neural mechanisms in the human brain. These experiments with the macaques are invasive, no doubt about it. Implanting the electrodes requires brain surgery. Although extreme care is taken to avoid discomfort in the implanted subjects, ethics preclude conducting such experiments on humans or the great apes (chimpanzees, gorillas, orangutans, and bonobos). The only exception to the rule is with certain neurological patients (most commonly epileptics) who have electrodes implanted for medical reasons. In such instances, single-cell research is perfectly ethical when permission is granted, as it almost always is. This limited research has yielded important results, as we will see later. And now, of course, the amazing new technology of noninvasive brain imaging (functional magnetic resonance imaging or fMRI, magnetoencephalography or MEG, and others that I will describe in later chapters) allows experimentation with human subjects that combines with the single-cell research on monkeys to yield the results and insights that are the subject of this book.

In setting the stage for the discovery of mirror neurons, I stated that the Parma investigators had acquired a pretty good picture of the actions of these motor cells during various "grasping" exercises with the monkeys. Let's now consider those early results in more detail. They are indeed fascinating, beginning with the fact that the motor cells fire *throughout the whole grasping action*, and not in correspondence with the contractions of any specific muscles. Even more surprisingly, the same cell often fires for both right-hand and left-hand actions and also, as anticipated earlier, when the monkey is moving the mouth. The team expected more specificity in the firing pattern—right hand only, left hand only, mouth only. What they saw, however, was this kind of specificity with the *type* of grasp the monkey was using. Some of these neurons fired only while the monkey was grasping small objects using two fingers, such as the handle of a cup using the thumb and the index finger. We call this type of grasp the precision grip. Other neurons in F5 fired only while the monkey was grasping large objects, such as a cup, using the whole hand—the whole-hand grip. It is in some way irrelevant to the monkey whether she grabs the cup with the right hand or the left hand, but the manner in which she grabs it is relevant. This is strange to us. Equally strange to us is the fact that these "grasping" brain cells do *not* fire when the monkey is scratching her head or performing another hand action, even though the very same finger muscles are used. These peculiarities suggest the existence of a somewhat complex vocabulary, in neural terms, of simple object-oriented actions and—the key point—at the level of the single cell.[7]

Of course, some—just some—of these cells also respond to visual stimulation, the surprising capacity that makes them either canonical neurons or mirror neurons. As discussed, the canonical neurons fire at the sight of certain graspable objects, the mirror neurons at the sight of grasping actions. As we might guess by now, these responses also have their peculiarities. The canonical neurons are sensitive to the *size* of the graspable object. For instance, if a cell fires when the monkey grasps a small object, such as a piece of apple, using the precision grip of thumb and index finger, the same cell will fire only when the monkey sees a comparably small object. This cell will not fire when the monkey sees a whole apple, which can be grasped only with a whole-hand grip. Likewise, canonical neurons in area F5 that fire when the monkey grasps a whole apple with a whole-hand grip will also fire when the monkey sees a whole apple, but will not fire when the monkey sees a raisin, which would require a precision grip to grasp. The correlation between action and perception in canonical neurons is tight indeed.

What about mirror neurons? Some mirror neurons also show this tight correlation between action and perception. These cells are called strictly congruent mirror neurons because they fire for identical actions, either performed or observed. For instance, a strictly congruent mirror neuron fires when the monkey grasps with a precision grip and when the monkey sees somebody else grasping with a precision grip. Another strictly congruent mirror neuron would fire when the monkey grasps with a whole-hand grip and when the monkey sees somebody else grasping with a whole-hand grip. Other mirror neurons, however, show a less strict relationship between executed and observed actions. These are the broadly congruent mirror neurons. They fire at the sight of an action that is not necessarily identical to the executed action but achieves a similar goal. For instance, a broadly congruent mirror neuron may fire when the monkey is grasping food with the hand and when the monkey is seeing somebody else getting food with the mouth.

In no case so far observed has the discharge of mirror neurons during action observation been modulated by the identity of the object to be grasped. Apple or orange? Peanut or raisin? It doesn't matter. Only the size matters, which makes perfect sense for motor purposes. Larger objects require the whole-hand grasp, smaller ones the precision grip. The discharge of mirror neurons during an observed action is also largely unaffected by the *distance* to the action. The scene can be close or far away. Mirror neurons also fire equivalently at the sight of a grasping human hand or a grasping monkey hand. They also fire similarly whether the experimenter grasping a piece of food eventually gives the food to a second monkey in the lab or to the subject monkey with the implanted electrodes. In short, the rewarding value of the grasping action does *not* affect the response of mirror neurons.[8]

A very interesting class of mirror neurons codes observed actions that are preparatory or logically related to the executed actions. A "logically related" mirror neuron is one that, for instance, fires at the sight of food being place on the table and also while the monkey grasps the piece of food and brings it to the mouth.[9] This class of cells may be part of neuronal chains of mirror cells that are important for coding not simply the observed action but also the intention associated with it. This intention is achieved through a sequence of simpler actions: reaching for the cup, grasping it, bringing it to the mouth, and then drinking from it.

A really telltale feature of the macaques' mirror neurons is that they do not fire at the sight of a pantomime. Performing a grasping action in the absence of an object does not trigger a discharge. This may seem odd to us, but it isn't particularly so, because these monkeys do not typically pantomime. We humans, however, do pantomime, and indeed our mirror neuron areas are activated by more abstract actions than are those of the monkeys. The several evolutionary steps dividing monkeys from humans can easily account for such difference. A subject for future discussion here will be the theory by computational neuroscientist Michael Arbib that mirror neurons are key precursors of neural systems for language. He proposes that pantomime plays a critical role in the evolutionary progression from the relatively simple mirror neuron system in monkeys to the much more sophisticated neural system that supports the high level of abstraction in human language.[10]

As we have seen, mirror neurons in area F5 fire at the sight of mouth actions as well as hand actions. In this capacity, these cells belong to two main categories: those that code for ingestive movements—eating a banana, drinking juice—and others that code for communicative movements such as lip smacking, a slight protrusion of the lips.[11] The existence of mirror neurons for communicative mouth movements suggested to Rizzolatti and his colleagues in Parma that these cells may have a profound role in the ability to communicate between individuals and in the understanding of other people's behavior. Thus they devoted a whole series of experiments to deeper aspects of the role of mirror neurons in coding the actions of other individuals.

I KNOW WHAT YOU ARE DOING

I am cooking dinner, and my daughter, Caterina, a sixth grader, is doing her homework on the table in the kitchen nook. I can watch her while I cook. The table is covered with books, notebooks, pencils, erasers, and so on. (I often feel that sixth graders these days have more homework than I had in high school.) As I cook dinner, I can't fully see what Caterina is doing. Her study materials are blocking my view. Still, I never feel I have to go on an elaborate inferential process to discern what she is doing. How is it possible? How can I have an immediate understanding of her movements even though I can't see them fully? Do my mirror neurons assist me in knowing and understanding what I cannot see? Alessandra Umiltà, now on the faculty at the University of Parma, was a graduate student in Giacomo Rizzolatti's lab when she led an experiment that tested this exact hypothesis.

The first two conditions of her experiment had been previously tested. In one, a monkey observed a human experimenter grasping an object. As expected, mirror neurons did fire at the sight of this grasping action. In the other condition, the monkey observed a human experimenter pantomiming a grasping action, in the absence of a real object to be grasped. As expected, the pantomime did not trigger a discharge in the neurons. With these standard but necessary results in hand, Alessandra added two new conditions to test whether mirror neurons would fire during an action the monkey cannot really see. In one, a three-dimensional object, for instance an orange, was positioned on the table. Subsequently, a screen was placed in front of the orange (other kinds of objects were also used, as these experiments typically involve several trials for each condition). With this screen occluding the monkey's sight of the orange, a human experimenter reached with her right hand behind the screen. The monkey saw the reaching but not the actual grasping of the orange. The question is: Did mirror neurons fire when the grasping action itself was hidden? The answer was yes (and no). Approximately 50 percent of the mirror neurons recorded in that experiment did fire; half did not.

In the other new condition, the table was bare. The screen was moved into position blocking the monkey's view of the bare table. Again, a human experimenter reached behind the screen with her right hand. Note that from the monkey's visual standpoint the experimental condition at this moment was identical to the previous one: the monkey was seeing a hand reach behind a screen. The only difference between the two conditions is the animal's *prior knowledge* concerning the presence of an object on the table. The question is: Did the monkey understand that this was a pantomime? If so, mirror neurons should not fire—and indeed they did not. The prior knowledge that there was no object on the table was sufficient for the mirror cells to now consider the hidden grasping action simply a pantomime, and thus not worth firing for.[12]

These experiments clearly show that mirror neurons do not simply form a neural system matching performed actions and observed ones. Even in the monkey, they provide a more nuanced coding of the actions of others, using prior information to differentiate the meaning of partially blocked actions that are visually identical. Is this sufficient evidence for us to conclude that mirror neurons code the intentions of the person grasping the object? Probably not, since the basic issue in the experiment was whether the hand grasped or not, as determined by the presence or absence of the object (the orange, in the example above). This experiment does not fully address the fundamental question of whether mirror neurons can differentiate between, say, grasping the orange in order to eat it versus grasping the orange in order to place it in the refrigerator. This is why Leo Fogassi, some years after Alessandra Umiltà's experiment, led another experiment that more explicitly investigated the role of mirror neurons in understanding intentions.

I KNOW WHAT YOU ARE THINKING

I am having a row with my wife over some family plans. We are in the kitchen, and she reaches for a glass. Does she want to drink or put it in the dishwasher—or maybe throw it at me? It is very useful to be able to predict what other people are going to do next.

The most basic property of mirror neurons—that is, firing for both the action of grasping a cup and the equivalent grasping action I only observe—suggests that they are helpful for recognizing the actions of other people. They also suggest that the action recognition process thereby implemented is some sort of simulation or internal imitation of the observed actions. Given that our own actions are almost invariably associated with specific intentions, the activation in my brain of the same neurons I use to perform my own actions when I see other people performing these actions may also allow me to understand the intentions of the other people. However, it can't be this simple. There's a problem, and it's the one I am facing when I see my wife grasping the glass during our row: the same action can be associated with different intentions. In fact, there is rarely, if ever, a one-to-one correlation: this action, this necessary intention. Just as I may have different intentions when I grasp a glass, so may others. Do mirror neurons differentiate between the same action associated with different intentions?

Leo Fogassi's recent experiment addressed this question directly by assessing the monkeys' neural activity during a variety of conditions of grasping *execution* and grasping *observation*. An explanation of the experiment requires a good deal of detail, but it is important for understanding these neurons. In one of the execution conditions, the monkey, starting from a fixed position, reached for and grasped a piece of food, then brought the food to the mouth to eat it. In a subsequent condition, the monkey reached for an inedible object in the same location as the food in the previous experimental condition. The monkey placed this object in a container. In this condition, several trials were performed while the container was located close to the mouth of the animal, such that the arm and hand movements of the grasping-to-eat and grasping-to-place conditions were *closely matched*. The main question was whether mirror neurons fire differently when the same grasping action leads to eating food as opposed to placing an object in a container. Does the intention matter to these neurons? (Note also that after completing the placing-in-the-container trials, the monkeys were rewarded, so the amount of reward for grasping-to-eat and grasping-to-place was identical.)

Between a third and a fourth of the recorded neurons fired

equivalently during grasping-to-eat and grasping-to-place. The majority of the neurons, however, fired differently, with about 75 percent discharging more vigorously during the grasping actions that brought food to the mouth, about 25 percent more vigorously during the actions to place objects in containers. What are we to make of these numbers? Maybe the differential discharging—the preference for eating over placing—was due to the fact that in one condition the monkey was grasping *food*, whereas in the other condition the animal was grasping a less interesting and useful object for placement. To test for this possibility, the monkeys were tested under the condition of placing food. The results were the same as in the previous experiment. The majority of cells discharged preferentially during grasping for eating, a minority preferred grasping for placing, and the minority of cells that had shown no preference for eating or placing still showed no preference. Conclusion: the type of object grasped was irrelevant. The important issue for mirror neurons was eating versus placing. Most "preferred" eating.

With these results in hand, the experimenters then proceeded to test the monkeys as they merely observed the same experimental setups, with an experimenter seated in front of the monkey performing the same grasping actions as the monkey had previously performed—some for eating, some for placing. With a container present and visible to the monkey, the experimenter grasped the food and placed it in the container. With no container present, the experimenter grasped the food, brought it to the mouth, and ate it. Thus the presence of the container acted as a visual clue, allowing the monkey to predict the next movement of the experimenter. The empirical question was whether mirror neurons would register a distinction when the monkey was observing grasping-to-eat versus grasping-to-place. The results demonstrated that the intention of the experimenter *did* make a difference, and the pattern of neuronal firing during observation of this grasping closely mirrored the pattern of neuronal firing as the monkey executed the grasping actions. If a cell discharged more vigorously while the monkey was grasping the food in order to eat, that same cell discharged more vigorously while the monkey was observing the human experimenter grasping the food in order to eat. If a cell discharged more vigorously while the monkey was grasping the food in order to place it in the container, that same cell discharged more vigorously while the monkey was observing the human experimenter grasping the food in order to place it in the container. If a cell discharged equally while the monkey was grasping-to-eat and grasping-to-place, that same cell discharged equally also while observing the human experimenter.[13]

The results of Leo Fogassi's experiment demonstrate that the coding of the actions of other people provided by mirror neurons is much more sophisticated than initially thought. Although Vittorio Gallese and Alvin Goldman had speculated soon after the discovery of mirror neurons that these cells may provide a key neural mechanism for understanding the mental states of others, they were in the minority at the time. Before Fogassi's experiment, there was much more support in the scientific community for a more parsimonious account of the functions of mirror neurons, with the cells simply providing a form of action recognition. Fogassi's experiment clearly supports the initial intuition of Gallese and Goldman. Mirror neurons let us understand the intentions of other people.

As I have stated, intention had always been considered off-limits for empirical study, too "mental." Not anymore. Fogassi's study and a soon-to-be-discussed imaging experiment with humans performed in my lab at UCLA strongly support the hypothesis that we understand the mental states of others by simulating them in our brain, and we achieve this end by way of mirror neurons. As I stated earlier, the fact that mirror neurons code differently the same grasping action associated with different intentions—not only when we perform the action, but also when we observe it in others—suggests that our brains are capable of mirroring the deepest aspects of the minds of others, even at the fine-grained level of a single cell.

I CAN HEAR WHAT YOU ARE DOING

I am working in my studio when I hear a distinctive noise from the living room. My daughter Caterina studies ballet, and she is at the age and proficiency level in which ballerinas start working en pointe. She is very excited and brings her new pointe shoes home for extra practice. Her steps with the pointe shoes make a distinctive sound on the hardwood floors.

By simply listening, I know what she is doing. I know a lot of things by simply listening. Clapping, tearing paper, typing, breaking peanuts—these are all actions that produce sounds and can be easily recognized by all of us. We don't give this ability a second thought. Most people think "we do it, that's all," but neuroscientists always ask *how*. And of course neuroscientists familiar with mirror neurons wonder whether these neurons might play a role in helping us recognize actions simply from hearing them. Evelyne Kohler and Christian Keysers are two such investigators. They performed their experiments on this subject in Giacomo Rizzolatti's lab.

Following the usual procedures, Kohler and Keysers identified mirror neurons in area F5 by measuring the responses of the cells while the monkeys were performing goal-oriented actions and then simply watching experimenters perform the same actions. The key, clearly, was that these actions—breaking a peanut, ripping a sheet of paper, and so on—produce sound. (As a control, the monkeys were also tested for white noise and other sounds unrelated to the actions. The control sounds were used to rule out the possibility that mirror neuron responses to action sounds were simply due to the arousing, nonspecific effect of any sound.) With all the necessary groundwork in place, Kohler and Keysers then recorded mirror neuron responses under three different experimental conditions: vision and sound, vision only, and sound only. For the "vision only" condition, objects were prepared so that they could be manipulated to perform an action visually similar to the naturally occurring action but without producing the sound. For instance, the peanuts were already broken in two parts and merely held in the initial position as if they were intact. The paper that was ripped was wet, so it produced no sound. In the "sound only" condition, digitized action sounds were used. There was no visual stimulation at all.

The results were clear and definitive: the mirror neurons discharged to all three experimental conditions. Some did seem to respond slightly more vigorously for the "vision and sound" condition, but the "vision only" and "sound only" conditions also yielded robust mirror neuron responses.[14] These results are very important because they demonstrate that mirror neurons code the actions of other people in a fairly complex, multimodal, and rather abstract way. Those cells that fire when the monkey herself is performing the sound-producing action also discharge at the mere sound resulting from somebody else's actions. That is, when we perceive the sound of a peanut being broken, we also activate in our brain the motor plan necessary to break the peanut ourselves, as if the only way we can actually recognize that sound is by simulating or internally imitating in our own brain the action that produces that sound.

Furthermore, the response of mirror neurons to auditory input is critical evidence in support of the hypothesized evolutionary link between these brain cells and language, the hypothesis proposed shortly after the discovery of mirror neurons by Giacomo Rizzolatti and Michael Arbib in a paper titled "Language Within Our Grasp."[15] The argument that mirror neurons are the evolutionary precursors of neural elements that enable human language was based first of all on an anatomical observation: area F5 of the monkey brain, where mirror neurons were recorded for the first time, is a homologue (that is, it corresponds anatomically) to an area of the human brain called Broca's area. Broca's area is an important brain center for language, named after the nineteenth-century French neurologist who discovered that a lesion here is typically associated with a disorder (Broca's aphasia) that affects mostly language production.

The argument for mirror neurons as language precursors also stems from the subtle consideration that these cells, by coding both for your action and your observance of that action in others, seem to create a sort of common code—and therefore a sort of "parity"—between you and the other individual. Several years before mirror neurons were discovered, Alvin Liberman had proposed that, since sending and receiving a message require, respectively, production and perception, the two processes of production and perception must somehow be linked and have, at some point, the same format.[16] Mirror neurons seem to provide precisely this common format.

However, the hypothesis that mirror neurons are precursors of neural systems dedicated to language had to face a problem. After all, language was initially spoken, occurring only through the auditory modality, whereas the sensory responses of mirror neurons had been initially investigated only in the visual domain. The discovery by Evelyne Kohler and Christian Keysers that mirror neurons also respond to action *sounds* provides strong support to the hypothesized links between mirror neurons and language. We will explore the language questions in much more detail in chapter 3.

MIRRORING TOOL USE

Until quite recently it was believed that only we humans use tools. We now know this is not true. Chimpanzees show some ability in using tools—nothing like ours, of course, but real nevertheless, and sufficient for scientists to study the evolutionary progression of tool use. In different locations in Africa, chimpanzees use the same tool, a stick, to achieve the same goal—eating ants—but they use this stick in fundamentally different ways from area to area. In the absence of any apparent difference in the environments of the different populations, the cultural difference suggests that the way tool use is learned and transmitted between chimpanzees is mostly through observation and then imitation.[17] Is it possible that mirror neurons are the brain cells that allow such learning via imitation?

In macaque monkeys, as we have seen, mirror neurons do not respond to the sight of a pantomime of an action. This makes sense because mirror neurons seem to code only for those actions that the monkey can perform—that are in its motor repertoire, as we say—and monkeys do not pantomime. By extension, mirror neurons in monkeys should have only a limited role in observational learning in general and in tool-use learning in particular, because monkeys are not so skilled at using tools. Take the Japanese monkeys who "wash pota-toes," a behavior that apparently spread from one precocious individual to the whole community. This famous case has spawned a significant debate in animal behavior literature. Initially, the behavior was taken as evidence that monkeys can imitate novel actions, but it was then argued that this may not fit a stringent definition of imitative learning. Ac-cording to the tougher standard, imitative learning requires learning a novel movement for your motor repertoire by watching somebody else performing the movement. A possi-ble explanation of the monkeys' behavior is that while the first monkey washes the potatoes, the attention of the observ-ing monkeys is directed to the *water* (this is called stimulus enhancement). The next time the observing monkey is close to the water with a potato in her hand, a simple trial-and-error mechanism during the manipulation of the potato in the water may have helped the animal to learn how to wash the potato. This would not constitute imitative learning, which is of a higher order. One fact in favor of this more conservative explanation is that the practice of washing potatoes did not spread as rapidly as one would have expected. This case and similar ones have provoked a variety of opinions within the community of scientists studying animal behavior, but it is fair to say that the majority of scientists do not consider washing potatoes as strong evidence in favor of imitative learning in Japanese monkeys.

If the washing-potatoes behavior did spread mostly through stimulus enhancement rather than imitative learn-ing, then it is very unlikely that mirror neurons played a critical role, since these cells respond to the observation of action. Attending to inanimate things such as water is not within their purview. Yes, mirror neurons must have been involved in the recognition of manipulating and holding the potato, but their role in spreading this behavior would be limited to this initial form of action recognition. If, on the other hand, the spread of the behavior is attributed to some form of imitative learning, then one might consider a more direct involvement of mirror neurons. This hypothesis also requires some evidence that mirror neurons may actually respond to the observation of some actions that are *not yet* in the motor repertoire of the monkeys. This evidence has been provided by Pier Francesco Ferrari, an ethologist who had studied animal behavior (especially forms of social contagion in monkeys) for years before training as a neurophysiologist in the lab of Giacomo Rizzolatti. Here is what he found.

Rizzolatti's previous finding was that mirror neurons that fire at the sight of a human experimenter grasping a raisin with a precision grip (thumb and the index finger) do not fire when the experimenter grasps the raisin with a tool, pliers for instance. This might seem odd on first consideration, but recall the hypothesis that mirror neurons do not fire at the sight of actions that are not part of the monkey's motor repertoire (thus the disinterest of these cells in pantomime, because monkeys don't pantomime). Likewise, monkeys do not naturally use tools, so the monkey's mirror neurons draw a critical difference between the precision grip and holding a pair of pliers.

Ferrari and his colleagues were recording neurons mostly

from the lateral part of area F5, a sector previously investigated only with regard to its motor properties, with the majority of cells coding for mouth actions. Ferrari's work came up with much more specific data. Almost all of these lateral F5 cells had motor properties, but there was a strong division of labor. Approximately one-fourth discharged during hand movements only, one-fourth during mouth movements only, and one-half during hand and mouth movements. Approximately two-thirds of the cells responded to visual stimuli; the majority were mirror neurons responding to the observation of actions of the experimenters. The novel finding, however, was that a robust contingent (approximately 20 percent of all recorded cells) responded to the observation of actions *performed with tools* (a pair of pliers and a stick with a metal tip). These mirror neurons responding to tool use also responded to actions performed with the hands and the mouth, but much more weakly. The main interest of this 20 percent minority was tool use.[18]

This discovery of a contingent of mirror neurons responding to the observation of "tool-use" actions is theoretically very important. These monkeys tested by Ferrari did not use tools themselves, so this is the first evidence of mirror neurons that prefer actions that are *not* in the motor repertoire of the observing animal. How should we interpret these findings? The first concept that comes to mind is that mirror neurons are concerned with *goals* more than with the specific actions to achieve those goals—a point demonstrated in the previously discussed data on the role of mirror neurons in distinguishing between intentions. The goal is the same whether the peanut is broken with the hand or with pliers. The goal is the same whether the food is grasped with the hand and eaten or speared with a stick and eaten.

This interpretation is plausible, but it does not explain why it took investigators so long to find neurons responding to tool use—approximately ten years from the first observation on mirror neurons. The Parma group had repeatedly attempted to measure such response but without success. I therefore think it is likely that these 20 percent of the mirror neurons in lateral F5 are the result of the repeated exposure of the animals to the sight of human experimenters using tools. This explanation of Ferrari's findings suggests that mirror neurons can acquire new properties, a key feature to support imitative learning. The formation of mirror neurons responding to tool use may be the first neural step in the monkey's brain to subsequently acquiring the motor skill to use those very same tools. Mirror neurons that respond to tool use are enticing evidence linking mirror neurons to imitative behavior, a powerful mechanism for learning.

I KNOW THAT YOU ARE COPYING ME

The case of the Japanese monkeys who wash their potatoes is just one example of an interest in imitation among animals that dates back at least to Darwin, who left detailed descriptions of various forms of mimicry in honeybees. There is ongoing heated debate on the question, as the old paradigm has been called into question. Among the naturalists of the nineteenth century there was a general consensus that imitation was quite widespread. For instance, George Romanes's book on animal intelligence, one of the most famous ethology treatises of the latter nineteenth century, describes monkeys as constant imitators: "they carry on this principle to ludicrous length." At the time, imitation was not considered the expression of a particularly high form of intelligence. Now it is. Indeed, a recent collection of essays describes imitation as "a rare ability that is fundamentally linked to characteristically human forms of intelligence, in particular to language, culture, and the ability to understand other minds." What a sea change since Romanes's times![19] And with this sea change comes another: the caution of researchers to acknowledge imitation as such. Behavior previously considered imitative in monkeys is now typically explained with some other, "simpler" cognitive mechanisms (such as the supposed stimulus-enhancement mechanism that might explain the spreading of potato washing in Japanese monkeys). This is now the dominant view among experts, but they must still deal with hard-to-refute evidence for imitative behavior in monkeys. Even neonates of rhesus monkeys are able to imitate some facial and hand gestures, such as lip smacking, tongue protrusion, mouth and hand opening, and opening and closing of the eyes.[20] Still, most scholars consider true imitation—that is, the ability to learn simply from observation—limited to humans and maybe the great apes.

This debate goes to the heart of the essential question underlying all research on macaque monkeys: Why do they have mirror neurons? The replies vary. Some researchers say that the real role of mirror neurons in monkeys is for action *recognition*, not action imitation. By activating mirror neurons in their own brains, observing monkeys recognize the actions of other individuals and apparently, to judge by the data from Leo Fogassi's experiments on intention, the goal of those actions as well. This is obviously a very important mechanism that facilitates social behavior in monkeys. However, other scientists—and I am definitely one of them—point out that there is some evidence, albeit not overwhelming, for true imitation in monkeys.[21] And even if one wants to discard such evidence, mirror neurons may also be involved in various forms of "contagion" (a technical term that does not imply disease). For instance, even assuming that the true mechanism of the spreading of potato washing among Japanese monkeys is stimulus enhancement, mirror neurons may be critical in the process by helping the recognition of the manipulative hand actions of the observed monkey. As it happens, Giacomo Rizzolatti, whose intuitions are to my mind without parallel in this field, has been quite conservative on these questions, emphasizing the role for these neurons in action recognition only. In recent years, however, he has been considering broader roles, and he is convinced of the role of mirror neurons in coding intentions. (As I've noted, his colleague Leo Fogassi's grasping-to-eat versus grasping-to-place experiments have had a powerful impact on the entire neuroscience community.) I believe that Giacomo is now more receptive to the idea that monkeys do imitate and that mirror neurons would be critical for such imitation.

Of course, imitation can work both ways, and a recent behavioral study in monkeys has provided inferential evidence that mirror neurons play an important role in the ability to figure out if another somebody is *imitating you*. Here, the experimenters adapted a paradigm devised by the developmental psychologist Andrew Meltzoff, an expert in imitation and social cognition in babies and young children. During the initial experimental phase, called the baseline period, the monkeys observed two experimenters, each manipulating a wooden cube with a hole in each side. The experimenters were mimicking typical actions the monkey would direct at the cube, such as biting, poking at the holes, and so on. Then a third cube was placed within reach of the monkey. When the monkey started manipulating the cube, one of the two experimenters imitated accurately the monkey's actions directed at the cube. The second experimenter, in contrast, performed different actions. For instance, if the monkey was poking at a hole, the imitator would also poke, while the non-imitator would bite. The monkey's behavior was taped and analyzed with intriguing results. Initially, the monkey did not show any visual preference between experimenters, but then the monkey clearly looked much longer at the imitating experimenter.[22] Clearly, the monkey—with what researchers call an "implicit" level of understanding—was able to recognize that one of the two humans was imitating her. An animal with an "explicit" understanding that she is being imitated will typically demonstrate behavioral strategies to test the imitator, such as sudden changes in behavior while looking directly at the imitator to gauge the response. The monkeys involved in the Ferrari study did not show this behavior. Their recognition that they were being imitated was understood only implicitly, but even this more limited understanding has an important social value.

The classic single-cell experiments on mirror neurons that might corroborate these behavioral results have not yet been conducted. They will be, and it is very likely that they will corroborate the behavioral study. And with that not-so-bold prediction, I close this chapter on the experimentation on mirror neurons of the macaque monkeys at the level of the single cell—vitally important work, not only because of its inferential value for thinking about our own brains (which we generally cannot access at that level, for ethical reasons), but also because it tells us where to aim the new, noninvasive technology with which we *can* study the mirror neuron system in humans. I now turn to these machines and this fascinating research, which confirms in every way the importance of mirror neurons for our experience as complex social creatures.

TWO

Simon Says

When people are free to do as they please,
they usually imitate each other.
—ERIC HOFFER[1]

COPYCAT CELLS

Imitation is not restricted to games such as Simon Says. Our drive to imitate seems to be powerfully present at birth and never declines. Gone is the nineteenth-century perspective that imitation is present just about everywhere in the animal kingdom, all the way "down" to Darwin's honeybees. In fact, imitation is deemed such a pervasive feature of human behavior that several authors have built theories placing it front and center. The most outspoken is probably Susan Blackmore, who argues in her book *The Meme Machine* that what fundamentally distinguishes humans from all other animals is not really language, the usual candidate for this accolade, but the ability to imitate.

In the 1970s, the American psychologist Andrew Meltzoff initiated a revolution of sorts in developmental psychology when he demonstrated that newborns instinctively imitate some rudimentary manual and facial gestures. The youngest of the babies tested by Meltzoff was all of forty-one *minutes* old. Every second of his life had been documented in order to demonstrate that he had not previously seen the gestures that Meltzoff performed for his experiment. Still, the baby managed to imitate those gestures. Thus, Meltzoff argued, an innate mechanism must be present in the newborn's brain that allows such rudimentary imitative behavior. This evidence was revolutionary because dogma held that babies learn to imitate in the second year of life, a belief originating in the work of Jean Piaget, probably the most influential figure ever in the field of developmental psychology. In effect, the Piaget school implicitly suggested that babies learn to imitate, but Meltzoff's data suggested that they may actually learn *by imitating.*[2]

This is an enormous difference, of course, and the hypothesis that we learn by imitating presents those of us who work with mirror neurons a wonderful opportunity to fill the explanatory gap. Given that newborns' brains do not have highly sophisticated cognitive skills, the fact that they can imitate suggests that the mechanism for this imitation depends on relatively simple neural mechanisms. When Meltzoff made his discoveries in the 1970s, mirror neurons had not been discovered—not in the macaque brain, not in the human brain—and would not be for another fifteen years. As we learned more about these neurons, some of us expected that they might be involved in early imitation in babies, but how would we gather the brain data required to confirm this hypothesis? Our imaging machines require that the subjects lie very still: it is not easy to convince babies to do that.

Recall the experiment discussed in the first chapter in which the monkey playing with the cube demonstrated more interest in the experimenter who was imitating the monkey than in the experimenter who was performing another action. We call this an "implicit" level of understanding, and we should not be surprised to learn that human babies also express it. Babies are delighted when a grown-up imitates them. If I am attending a gathering with friends and a young baby is in the house, the first thing I do is imitate what the baby does. Suddenly I become the most popular grown-up in the eyes of the child (excluding the parents, of course). Babies love playing imitation games. Obviously, there is also plenty of reciprocal imitation between parents and babies. In fact, this specific imitation (and chemistry) may be one of the major shaping factors in reinforcing mirror neurons in the developing brain. I will discuss this hypothesis in a later chapter.

There is also plenty of reciprocal imitation among toddlers. The developmental psychologist Jacqueline Nadel facilitates this form of spontaneous imitation by setting up a playroom with two of everything. It is amazing to watch the spontaneous imitation in these settings among very young children who have not yet developed language. When one child puts on a hat, the other child puts on the second hat; when the first child adds sunglasses to the outfit, the other follows suit. When one picks up an umbrella, the other child picks up the other umbrella. When the first child starts spinning the umbrella, the second child spins hers too. Down goes one umbrella, down goes the other umbrella; grab a balloon, grab a balloon. There's no end to the imitation game. Even the little hand-waving gestures of one child holding the balloon are imitated by the other child. Another developmental psychologist, Carol Eckerman, has shown strong ties between imitation and verbal communication in children. When toddlers who do not know how to speak interact, they tend to play imitation games. The more a toddler plays imitation games, the more the same child will be a fluent speaker a year or two later. Imitation seems like the prelude and the facilitator of verbal communication among young children.[3]

As adults, we do not lose our childhood fondness for and employment of imitation. On the contrary, imitative behavior is strongly present in adulthood. By transmitting social practices from generation to generation, it has produced the vast range of different cultures throughout the world. It has also produced thousands of languages over tens of thousands of years, and it is still producing regional accents as we speak.

In *The Meme Machine*, Susan Blackmore adopted the key word in the title from Richard Dawkins. (Should I say she imitated him?) For his part, Dawkins was well aware of the power of imitation in transmitting mannerisms, practices, ideas, and even entire belief systems, and he coined the term "meme" some thirty years ago in his famous book *The Selfish Gene*. His idea was to "imitate" (again!) or borrow concepts from biology and genetics by creating an analogy between the transmission of genes down the generations with the transmission of behaviors down the same generations. His term (that is, his meme) has become so successful that it is now included in the Oxford English Dictionary with the following definition: "An element of a culture that may be considered to be passed on by non-genetic means, especially imitation."

As the word "meme" imitates the word "gene" to signify transmission of information in the behavioral domain, so the word "memetics" now describes a whole approach to evolutionary models based on imitative transmission.[4] Indeed, memes and memetics have inspired many authors and spawned a host of ideas in the field of evolution, even in the philosophy of mind.[5] Daniel Dennett, for one, proposed in *Consciousness Explained* that memes played a major role in the evolution of human consciousness. Indeed, Dennett sees consciousness as the product of the interactions between memes and brains. He writes, "The haven all memes depend on reaching is the human mind, but a human mind is itself an artifact created when memes restructure a human brain in order to make it a better habitat for memes." According to Susan Blackmore, memes "are the very stuff of our minds. Our memes are who we are." This concept may sound dangerously close to limiting our notion of free will. If so, it will not be the last one in this book to do so. As I discuss later, our research on mirror neurons suggests that our notion of free will may have to be revised.

One aspect of memes that Blackmore and others emphasize, following the selfishness of genes proposed by Richard Dawkins, is the ability of memes to replicate themselves by a process of "infecting" a large number of brains. A good example of very active memes are the ubiquitous "urban myths." Ironically enough, one such robust urban myth—now actually an international myth—involves the discovery of mirror neurons themselves. Remember the uncertainty about the very first serendipitous observations of mirror neurons in Giacomo Rizzolatti's laboratory in Parma? One of the numerous stories going around the world of science holds that Vittorio Gallese was licking an ice-cream cone in the lab when a neuron wired in the macaque's brain started firing. I heard this story several times in several places, and at some point I started believing it myself. In fact, I became one of the vehicles for transmitting this meme, because I told the ice-cream story in seminars and even to some journalists interviewing me about mirror neurons. I planned to incorporate it in this very book, but I decided that I should first ask Rizzolatti and Gallese about its veracity. Alas, it turns out that the ice-cream story is not true at all. Nobody knows how started it or why, but it is charming and has proved to be an appealing and tenacious story, both to tell and to hear.

As they put together the implications of their puzzling discoveries in Parma, Rizzolatti and his colleagues were familiar with the theory of memes, and they realized when the various explanatory pieces fell into place that the properties of these heretofore unsuspected neurons fitted perfectly with that theory. These very specialized cells seemed to be the brain's ideal enablers of imitation as well as of other aspects of our social behavior. It was time to learn more by expanding the research on mirror neurons in monkeys to include studies in humans utilizing the marvelous, noninvasive, and costly brain imaging technologies now at scientists' disposal.

The new work would build on a small number of experiments with humans that had already produced very interesting results. One was several decades old. In it, two psychologists had obtained what might have been the very first experimental clue about mirror neurons in humans when they measured the muscle activity of subjects observing two individuals arm wrestling and, in a different setup, a stutterer reading. The two experimenters used electrodes to measure muscle activity in the subjects' foreheads, palms, lips, and arms as they observed the scene in front of them. In light of what we know today, the results are perfectly understandable. The muscle activity recorded in the subjects' lips was highest while they were watching the stutterer; activity was highest in the arms while subjects were watching the arm wrestlers![6] Just as objects in experimental physics tend to vibrate when excited by energy at specific frequencies, so the muscles of the observers seemed to resonate with the hardworking muscles of the active participants.

Then there were the "prototype" imaging experiments using positron-emission tomography (PET), a technique that uses radioactive material to measure blood-flow changes and other physiological parameters in the brain. In these studies, subjects were asked to grasp objects, watch other people grasping objects, imagine grasping objects, or simply watch graspable objects without grasping them. The first two of these conditions—grasping and watching—were closely analogous to the single-cell mirror neuron experiments with macaque monkeys, of course, and the results confirmed that two areas of the human brain, anatomically similar to those in the macaque brain that contain mirror neurons, were also active during both execution and observation of grasping actions, and even during imagination of grasping actions. This was already an exciting initial result.[7] It was joined by another early (mid-1990s) attempt to collect evidence in favor of mirror neurons in the human brain, one imagined and orchestrated by Luciano Fadiga, one of the neurophysiologists in Parma. This work, using transcranial magnetic stimulation (TMS), was so ingenious for the time that I want to explain it in some detail. It is also similar in a way to the "muscle resonance" experiment with the arm wrestlers many years earlier.

For the basic experiment, a specially designed copper coil wrapped in plastic is placed over the head of the intrepid subject. (I'm joking about the "intrepid." The setup sounds intimidating, but it was perfectly harmless.) Electrodes taped to the subject's right hand will register any activity in these muscles. With everything ready to go, the subject either watches an experimenter performing grasping actions or some activity completely unrelated to the hand—for instance, a dimming light. In each case, a mild electric pulse is simultaneously routed through the coil, thus creating a magnetic field that induces an electric current over the surface of the brain. This current, in turn, induces a baseline excitement in the cells of the primary motor cortex, which will then be further influenced by the observation of the chosen scene.

Fadiga's reasoning was this: If humans have mirror neurons, these cells are probably located in the premotor cortex, the brain sector important for motor planning and analogous to area F5 of the macaque brain containing mirror neurons. Since the premotor cortex is connected to the primary motor cortex, the excitation of mirror neurons in the premotor cortex while the subject observes the experimenter grasping an object should also make the primary motor cortex more excitable, resulting in signals sent to the hand muscles, which would twitch involuntarily. On the other hand, any mirror neurons would be less excited about the dimming light, and any resulting twitching in the hand would be less pronounced. Indeed, Fadiga and colleagues measured just this difference: watching grasping actions produced bigger muscular twitches than watching the dimming light. Furthermore, these bigger muscular twitches were measured only in the *specific hand muscles* that are involved in grasping, not in the many uninvolved muscles in the same hand. Even though the subjects were completely still, the motor system in their brains was quietly pretending to perform (or "simulate," as most scientists would say) the actions that the subjects were simply observing.[8]

The results of all three of these earlier experiments were completely in line with the work on the monkeys and quite convincing as demonstrations of the existence of a mirror neuron system in the human brain. Now, in the mid-1990s, the Parma group wanted to enlist all of the even newer technology in the service of this important research. To this end, Vittorio Gallese, the colleague who brought to bear on the research in Parma his understanding of Merleau-Ponty and phenomenology, organized and led an international research project funded by the Human Frontier Science Program Organization. This group funds international collaborative projects, bringing together labs from different countries, perhaps even different continents, as in this case. The idea is to study the same topic using different methodologies, at the same time fostering international collaborations that can provide cultural exchanges between labs.

Gallese's project included seven labs, five countries, and three continents. Three of the labs studied monkeys. The Italians in Parma continued their investigations using depth electrodes to measure single-neuron activity in the macaques. Another European group on the beautiful Mediterranean island of Crete investigated the monkeys' mirror neuron system by using brain imaging rather than depth electrodes. The third group working with monkeys was a Japanese team in Kyoto that had access to one of the largest colonies of monkeys in the whole world. They would build a sort of library, or database, of the monkeys' communicative facial expressions, which would stimulate future experiments on the monkey brain's responses to facial expressions.

Two other groups would study the mirror neuron system in humans. One, in Helsinki, Finland, used magnetoencephalography (MEG), and the other, at UCLA, of which I was and still am a member, used mainly functional magnetic resonance imaging (fMRI). The remaining two groups were, respectively, computational neuroscientists in Los Angeles, who worked out models of the mirror neuron system, and engineers in Cagliari, Italy, who implemented virtual reality environments used in experiments.

In the end, these seven labs funded by the Human Frontier Science Program Organization produced a series of experiments that greatly expanded our understanding of the mirror neuron systems in both monkeys and humans. The project also helped enlarge the group of scientists working in this new field. Just a decade ago, mirror neurons were known to very few researchers and largely ignored by others outside the core group. It is hard to believe that in just a decade they became the most "popular" brain cells of all.

RESONATING BODIES

There are several evolutionary steps from macaques, which are small apes, to chimpanzees, which are great apes, to the hominids who preceded humans (and are no longer around to be studied), and finally to humans. How did the mirror neuron system change through these many evolutionary steps? What kind of functions could the mirror neuron system support in humans that are not supported in macaques? These were the questions we wanted to address at UCLA.

Our first focus was imitation. Our hypothesis was that mirror neurons were involved in the evolutionary progression from monkeys—who have an implicit understanding of imitation, as we saw in chapter 1—to humans, who are utterly proficient imitators. To test this hypothesis, we collaborated with the group in Parma that discovered mirror neurons in monkeys and with Marcel Brass and Harold Bekkering, two members of the Max Planck Institute for Psychological Research in Munich, Germany. Brass and Bekkering had already performed several behavioral investigations into imitation in children and adults, work inspired by the *ideomotor model* of human action. This model differs substantially from the more popular *sensory-motor model*, according to which the starting point of human actions is some form of sensory stimulation, with actions coming into play only as a response to the initial stimulation. By contrast, the ideomotor model of human actions assumes that the starting point of actions are the *intentions* associated with them, and that actions should be mostly considered as means to achieve those intentions.[9]

In turn, the ideomotor model is rooted in the insights of two philosophers of the nineteenth century, the German Rudolph Hermann Lotze and the American William James, who independently elaborated on the idea in their discussions of voluntary actions and their consequences. Their main concept was that voluntary actions require a representation of what is going to be achieved, a representation that has to be unchallenged by a conflicting idea. When these two conditions are met, the representation of the intended goal is sufficient to directly activate the action. How does this happen? According to the ideomotor model, it happens because we humans have learned about the effects of our own actions. For instance, if you have learned in the past that pressing a specific key of your computer turns it on, just thinking about turning it on will activate in your brain the representation of the finger movement you use to press the key.[10] This concept, developed by two nineteenth-century philosophers, seems like a very good description of what mirror neurons do. Indeed, according to the same logic of the ideomotor model, watching somebody else's actions and their consequences should activate the representations of your own actions, which are typically associated with the same consequences (for instance, seeing someone else turn on her computer would activate the representation of your finger movement to turn on your computer).

At UCLA, we built on the previous work of Brass and Bekkering with the ideomotor paradigm for our first brain imaging experiment on the role of the human mirror neuron system in imitation. Our tool of choice for the experiment was fMRI, a large machine that uses a powerful magnet to set up a magnetic field. The way fMRI measures brain activity, specifically, is relatively simple. Suppose you want to wiggle the fingers of your right hand. In order to do that, cells in your motor cortex discharge action potentials that send electrical signals to the spinal cord and the muscles of your fingers. This neuronal firing costs energy. In order to supply the oxygen the brain cells need when they fire (pretty much as the engine of your car needs gasoline to keep running), cerebral blood carrying the oxyhemoglobin protein rushes into your motor cortex. The brain cells take up the oxygen from this protein, which then becomes deoxyhemoglobin—that is, hemoglobin without oxygen. For fMRI, the key fact is that oxyhemoglobin and deoxyhemoglobin have different magnetic properties and behave differently in the magnetic field created by the magnet of the MRI scanner. Moreover, the blood flow that rushes into an activated brain area—in this case, the motor cortex—is in excess of what is needed, so the proportion of oxyhemoglobin and deoxyhemoglobin in the blood changes when a given brain area is activated. An activated area has a higher proportion of oxyhemoglobin, so therefore the level of blood oxygenation is a good indicator of brain activity in a healthy brain. Thanks to the fortuitous combination of all these natural phenomena, it is possible to use fMRI noninvasively to track the activity of the whole brain while subjects are performing a variety of tasks. The fMRI is completely safe and allows the study of the whole brain simultaneously. It has good, although not ideal, spatial and temporal resolution—that is, it allows resolution to the level of a few millimeters of brain tissue, but not down to the single cell achieved with the implanted electrodes in experiments with monkeys, and it pinpoints events to about one second, if not the milliseconds of single-cell research. The combination of all these factors is what makes fMRI so successful in the neurosciences today.

With fMRI, the only "catch," really, is the absolute stillness required of the subject lying inside the machine. Head movements ruin everything, for obvious reasons. Tiny movements are tolerable—the software picks these up and makes compensations—but stillness is the rule. Also, the hollow tube is a confined space, with just a few inches between the face and the "wall," so there's no room for a standard computer monitor or projection screen. If the experiment requires (and it usually does) that the subjects observe some kind of scene or display play, subjects are outfitted with high-tech goggles featuring two very small high-resolution LCD monitors. Typically, the subject is on the job for about an hour—although not an hour without moving at all. There are a lot of breaks, during which one can move a little. Typically, the whole experiment, including prep work and analysis, will take several weeks, maybe two or three months. Of course, conceptualizing that experiment might have taken years of on-and-off pondering and noodling and consultation.

For our maiden imaging experiment on the role of the human mirror neuron system in imitation, the subjects performed, imitated, and watched certain hand movements. We reasoned that during imitation, our subjects were *also* observing and performing the imitated actions, by definition. Hence, we expected that brain areas with mirror neurons would have a level of activity approximately equal to the sum of the activity measured during action execution and action observation. (Before starting the experiment, we had to figure out how to translate the single-cell data recorded in monkeys with the expected pattern of activation in human mirror neuron areas. The data from Parma clearly demonstrated that the discharge of mirror neurons while the monkeys were performing grasping actions was approximately twice as strong as the discharge while they were simply observing somebody else grasping. We therefore expected that action execution was going to activate human mirror neuron areas approximately twice as much as action observation.) Indeed, we found two human brain areas with such activity, and they corresponded nicely to the anatomical locations of the two macaque brain areas (area F5 in the frontal lobe and area PF in the parietal lobe) in which mirror neurons had been recorded in Parma.

These anatomical correspondences, shown in figure 1, demonstrate the evolutionary continuity between the macaque and the human mirror neuron system. The brain of the macaque is much smaller than the human brain, with a less complex shape. However, there are many similarities between the

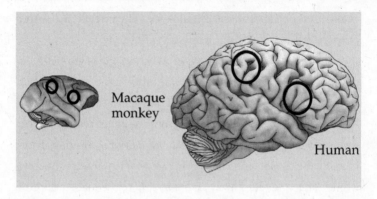

Figure 1: Brain areas with mirror neurons in macaques and humans.

two with regard to the major bumps (gyri) and grooves (sulci). These similarities make comparisons between the two species relatively easy. Both brains are divided into left and right hemispheres; this figure shows the right hemisphere, with the front of the brain to the right. The two well-identified, anatomically similar brain areas with mirror neurons in macaques and humans are located in the frontal lobe and in the parietal lobe behind it. Importantly, the left frontal lobe area with mirror neurons is Broca's area, a major human brain area for language, thus also supporting the evolutionary hypothesis that mirror neurons may be critical neural elements in the evolution of language.

DO WHAT I SAY, NOT WHAT I DO

Imitation shapes human behavior very powerfully. We all should remember this—particularly those of us who still have relatively young kids. I have noticed that people usually *say* the right thing to their kids: you should never lose your temper, always consider the point of view of other people, and so on. But the question is: Do we do what we say? In some cases I find myself modeling for my daughter exactly the behavior I tell her not to adopt! In these cases, I am afraid she will pick up mostly what I do, not what I tell her she should do, because the child's brain is extremely good at picking up behaviors from other people through imitation. (As the child grows older, such forms of imitation would be much more complex than the basic mimicries observed in very young infants, as described earlier, and more complex than the imitative behaviors in older children, as described in the following paragraphs. These "higher" and more complex forms of imitation are the subject of extensive discussion later.)

Even the young child seems highly attuned to pick up the *goal* of other people's behavior, as Harold Bekkering and his colleagues at the Max Planck Institute in Munich demonstrated in a clever experiment with preschoolers. In the experiment, children were simply told that they were going to play a game with a grown-up. When the game began, the instructions from the grown-up were simple: "Do what I do." Sometimes the adult touched the left or the right ear with the hand on the same side (the ipsilateral movement), sometimes with the hand on the opposite side (the contralateral movement). Sometimes the adult used both hands simultaneously to touch the ear on the same side, sometimes both hands to touch the ear on the opposite side. The children made mistakes, mostly when they were supposed to imitate a movement of the hand touching the opposite ear. In these cases, they often touched the correct ear but with the wrong hand—that is, the one on the same side as the ear.

Could it be that the children had problems in making a hand movement that would cross the midline of the body? No, because when they imitated movements with both hands, they basically made no mistakes, even when they were imitating the hand movements touching the opposite ears. Clearly, the children did not have problems in doing the hand movement reaching the opposite ear, yet when just one hand and one ear was involved, they made mistakes. What was going on? It was an interesting puzzle. The children's mistakes made sense only if they were prioritizing, creating a hierarchy of sorts, an estimation of what really counts when one imitates other people. And to the children it must have seemed as if what really counted was the *goal* of the hand movement, that is, the *specific ear* to be touched. Focused on that goal, they did touch the ear that was supposed to be touched, just with the wrong hand, probably because it was closer to their intended goal. When the adults were using both hands and touching both ears, children did not need to focus on the ear to be touched (indeed, both ears were touched) and could focus instead on the movements of the hands.

To further test the hypothesis that the key for the children was the apparent goal of the action, Bekkering and colleagues performed a second experiment. Here they limited the hand movements to touching only one ear. Some children were shown only touches directed to the left ear, other children only touches of the right ear. With this setup, Bekkering and colleagues reasoned, they had eliminated for the children the problem of choosing the ear. Without it, they might make the imitation of the touching hand the primary goal of the game. Indeed, this is what happened. When only one ear was "in play," the children imitated almost to perfection, using either hand as called for. The action in this setup—the contralateral movement of one hand across the midline of the body to touch the opposite ear—was identical to the action that had caused the confusion initially, but in this simpler version of the task, the goal of the action was always the same, helping the imitative performance of the children.

For one final test of the hypothesis, Bekkering and his colleagues studied another group of children, using yet another game. This time they wanted to find out whether the same hand movement can be imitated correctly depending on how the child represents the goal of the action. In this setup, the adult sat on one side of a table, the child on the opposite side. For the ipsilateral movement, the adult extended either the left or the right hand over the table and placed it on a specific location on the same side of the body. For the contralateral movement, the adult extended the hand over the table and placed it on the table in a specific location on the opposite side of the body. Again, the children were simply told, "Do what I do." And they did, successfully imitating both types of movements with both left and right hands. So far, so good. In the next session, however, two red dots were placed on the table, on the same spots where the adults had placed their hands in the previous session. Once again, the adults performed the hand movements. This time they placed the hands over the dots. Again instructed to "do what I do," the children started making mistakes when they were supposed to imitate the contralateral movements. Indeed, this mistake was the same mistake made by the children imitating the hand movements touching the ears in the very first experiment. Supposed to imitate contralateral movements, the children instead reached for and covered the correct dot, but with the wrong hand.

The explanation was now straightforward: The presence of the dot had modified the goal of this strange game they were playing. For the children's brains, the goal of the imitation game was now to "cover the dot," and the most direct way of doing it was to use the hand on the same side as the dot. In the imitation session without the dot, the hand movement itself was the goal of the observed action, which is why the children imitated perfectly that time.

These imitation games of Bekkering and colleagues show us that coding the goal of the observed action is the primary factor in driving imitative behavior in preschoolers. Such goal-oriented imitation is probably driven by mirror neuron activity—even in children. Bekkering and his colleagues then extended this experimental work to adult subjects, performing an experiment similar to the one with the dots on the table that had induced the mistakes in the preschoolers. Obviously—and thank goodness—the adults did not make mistakes, but here's the really interesting payoff: the analysis of the *response times* by the adults demonstrated a pattern similar to the pattern of mistakes made by the children—that is, a slight delay for the contralateral movement, thus showing that coding goals is the primary factor in adult imitation too.[11]

What do these experiments have to do with mirror neurons? Everything, because these neurons seem to be the ideal brain cells to implement this form of imitation rooted in imitating goals. As we have seen in the first chapter, even mirror neurons in the macaques seemed concerned much more with the goal of the action than with the action itself.

My group at UCLA, inspired by their imitation games with children, now teamed up with Bekkering and one of his colleagues, Andreas Wohlschläger, to perform a brain imaging experiment on adults. As in those games, our subjects saw and imitated ipsilateral and contralateral finger movements; in some cases the finger was placed on a red dot. Subjects were simply told to observe and imitate the index finger movements. No mention was made of the dots. Our hypothesis was that the brains of our subjects would code "cover the dot" as the goal of the action. Thus their mirror neurons, more concerned with goals than with mere actions, should be more active when the subjects imitated the finger movements that covered the dot. Indeed, this is what we found: the mirror neuron area in the frontal lobe corresponding to Broca's area had much higher activity while subjects were imitating finger movements covering the dots, compared with finger movements that did not cover the dots, even though the two movements were identical.[12]

We turned next to another form of imitation that also shows strong prevalence in young children, as confirmed almost forty years ago by the psychologists Seymour Wapner and Leonard Cirillo. In this study, one of the experimenters gave the usual instruction to a group of children: "Please do what I do." He then raised the right hand. The younger children in the experiment—say, first graders—would raise their left hand, imitating the adult as if they were in front of a mirror. In older children—kids in junior high school, say—this instinct for mirror imitation disappears. They raise the anatomically correct hand. At UCLA, we reasoned that if mirror neurons are critical brain cells for imitation—and especially in development—even in adults they should show the "preference" revealed *not* by the older junior high kids, but by the young children. Even in the adult brain, they should faithfully reflect the behavioral tendencies of little children. To find out, we set up a brain imaging experiment calling on the adult subjects to imitate finger movements, but with the right hand only. In some cases, however, they would be imitating with their right hand observed finger movements of the *left* hand—that is, they would be restricted to imitating as if in front of a mirror. We hypothesized that in these cases mirror neurons would be much more active. And they were. Indeed, they were activated four times more strongly during the mirror imitation as compared with the anatomically correct imitation, even though the finger movements were identical in both cases.[13]

Imitation in young children is both goal-oriented and performed "as in front of a mirror." This is certain. Now the question was how to bring together conceptually these two aspects of imitation. In short, what is the *goal* of imitating as if in front of a mirror? We can start with the observation that two people face-to-face and imitating each other as if in front of a mirror thereby use the same sector of space. When you and I are facing each other and imitating each other, my right hand is in the same sector of space as your left hand. We "share" this same space and thereby get literally closer to each other. I think one of the primary goals of imitation may actually be the facilitation of an embodied "intimacy" between the self and others during social relations. The tendency of imitation and mirror neurons to recapture such intimacy may represent a more primary, original form of intersubjectivity from which self and other are carved out. (Much more on this subject later.)

Indeed, studies on unconscious forms of imitation during naturalistic interactions, performed well before mirror neurons were discovered, support these concepts. In one study on postures, the psychologist Marianne LaFrance looked at arm and torso positions of students and teachers during regular class periods, classifying imitative postures as either mimicking (that is, using the anatomically correct arm, say, teacher's right arm, student's right arm) or mirroring (teacher's left arm, student's right arm). When LaFrance correlated the overall rapport in the classroom with the unconscious imitation of the teachers' postures, she found that the higher the rapport, the more *mirroring* as opposed to mimicking. In another study, perceived mirroring in face-to-face interactions was also judged to convey more solidarity, involvement, and "togetherness." Here, subjects were watching pairs of pictures depicting face-to-face interactions in which people are leaning either in opposite directions (both to their left side, say) or in the same direction (the left side of one, the right side of the other). Pictures of people leaning toward the same direction were found to convey more closeness than pictures of people leaning toward opposite directions.[14]

Once again, and even though it is practically impossible to obtain brain imaging data in such naturalistic contexts, it makes sense to hypothesize the involvement of mirror neurons in the spontaneous mirroring behavior of people, especially in light of our brain imaging data on goal-oriented and mirror imitation. The intimacy of self and other that imitation and mirror neurons facilitate may be the first step toward empathy, a building block of social cognition, as we will see in chapter 4. The study of early human development also shows how imitation is powerfully linked to the development of important social skills—for instance, the understanding that other people have their own thoughts, beliefs, and desires. If imitation is so critical to develop these social skills, mirror neurons that enable imitation must be too.[15] In the first chapter I discussed the single-cell data in monkeys showing that mirror neurons code the intention associated with the observed action. Now let's consider the brain imaging evidence in humans that supports the same conclusion.

HARRY POTTER AND PROFESSOR SNAPE

On January 10, 2006, the science writer Sandra Blakeslee published an article in *The New York Times* on mirror neurons. The title was "Cells That Read Minds." I guess that

Blakeslee or her editors wanted to emphasize one of the most astonishing implications of the discovery of mirror neurons. Their relatively simple physiological properties allow us to understand the mental states of other people, an ability that has always been perceived as somewhat elusive. The most popular existing expression for this ability is "read minds." I believe this expression is already charged with specific and incorrect assumptions about the process we are trying to understand. The term "mind reading" implicitly suggests that our understanding of the mental states of others requires the use of inferential or symbolic thinking. Indeed, this has been the dominant assumption among the scientists interested in the cognitive faculty of understanding the mental states of others.

According to the dominant view, we (starting when we are children) understand the mental states of other people using the same approach scientists use to understand natural phenomena. After observing other people's behavior, we concoct theories about their mental states, just as physicists would make theories about physical systems. Then we look for evidence in support of the theory. If the evidence does not support the theory, we revise the theory, perhaps even create a new one. For instance, if we see somebody crying after she falls, we theorize that the crying expresses pain. However, later on we may see someone crying while receiving a prestigious award, which forces us to revise our theory about crying and the mental and emotional states associated with it. This model of understanding the mental states of other people is called, in scientific jargon, "theory theory" (perhaps confusingly), because the understanding of mental states of other people is somewhat similar to a scientific theory: these states cannot be directly observed, but the behavior of others can be predicted on the basis of a set of causal laws bringing together perceptions, desires and beliefs, decisions and actions.

I have always thought that this model of how we understand the minds of other people is too complex and, not coincidentally, dangerously similar to the way the people who propose it (academics, of course) tend to think. I base my doubts about the theory theory on the straightforward observation that we understand the mental states of other people almost continuously, often without giving the problem too much thought. In my seminars, when I want to introduce the concept that nature may have devised a much simpler, much less laborious way of understanding the mental states of our fellows, I use a conversation between Harry Potter and Professor Severus Snape in *Harry Potter and the Order of the Phoenix*, the fifth book of the saga. (Like most parents, I guess, I started reading the series at the behest of my daughter, and I was soon addicted of my own accord.) In this scene, Lord Voldemort, a very mean wizard indeed, is trying to enter Harry's mind in order to acquire important information that he can use for his evil plans. Professor Snape is supposed to teach Harry the art of occlumency—that is, the ability to stop others from entering one's own mind.

"The Dark Lord is highly skilled at . . . extract[ing] feelings and memories from another person's mind," he says.

Harry is quite surprised and excited. "He can read minds?"

"You have no subtlety, Potter . . . Only Muggles talk of 'mind reading.' The mind is not a book."

Although I definitely do not like Snape, I must confess that his reply to Harry summarizes well my position on the argument of understanding other minds. The mind is not a book. I do not think we "read" other minds, and we should stop using terms that already contain bias about the way we think about such a process. We read the world, yes, but we do not read other minds in the usual sense that that phrase is used.

I do not believe we need to overload the brain with complex inferential thoughts on why people do what they do or what they are going to do next, especially for the more or less continuous understanding of the simple, everyday actions of our fellow humans. We have people around us all the time. We would not be able to cope with all this if we had to be scientists, like Einstein, analyzing every person around us. Nor was I the lone wolf in my opposition to the theory theory. When it was the dominant model in the field of developmental psychology—well before mirror neurons were discovered—a minority of scholars proposed an alternative called simulation theory. This one holds that we understand other mental states by literally pretending to be in other people's shoes. There are two variations on this idea, one more radical than the other. The moderate version maintains that the pretense to be in somebody else's shoes is a cognitive, deliberate, and effortful process, whereas in the more radical variation it is thought that we automatically simulate in a fairly unconscious way what other people do. On this question I am a radical, since this automatic, unconscious form of simulation maps well with what we know about mirror neurons.[16]

After the discovery of mirror neurons, the popularity of the theory theory as an explanation of how we understand other minds declined dramatically. Acceptance of the simulation account increased tremendously. However, the empirical data demonstrating that mirror neurons code the intentions associated with the observed actions was lacking until recently. The first such study was a collaborative experiment between my lab and the neurophysiologists in Parma, in particular Giacomo Rizzolatti and Vittorio Gallese. We had the idea a long time ago, while we were in Crete for one of our early meetings in the fall of 1999. The paper that reports our findings was published in 2005. It took us a while to design the experiment. Intentions are slippery things to be studied with empirical research. You will recognize this experiment: I mentioned it briefly in the first pages. We are back with the teacups.

GETTING A GRIP ON OTHER MINDS

The way we proceeded was conceptually similar to the "grasping-to-eat or grasping-to-place" experiment Leo Fogassi performed with monkeys, described in chapter 1. (Indeed, our imaging experiment actually preceded and inspired Fogassi's experiment.) Our initial idea was that the same action can be associated with different intentions. One can grasp a cup for many reasons. The two most common ones are probably to drink from it or to put it in the dishwasher. The context often gives clues to the observer regarding which intention is the most likely one. For instance, if we have just begun enjoying breakfast and I see my wife reaching for her teacup, it is likely that she will drink from it. However, if we have finished eating and she reaches for the cup while standing up, it is likely that she will put it in the dishwasher. True, she could be having a last sip. However, this outcome is less likely than putting the cup in the dishwasher, on the basis of the context in which my wife's action is embedded.

If mirror neurons respond only to the grasping action, it does not really matter which context surrounds them. Indeed, it does not really matter whether there is any context at all. A grasp is a grasp, with or without context. If, on the other hand, mirror neurons respond to the intention associated with the observed action, as they do in monkeys, then the context should influence the activity of mirror neurons. Following this logic, we designed a brain imaging experiment in which subjects saw a series of video clips. One type, which we called Action, simply displayed a hand grasping a cup in the absence of any context. Different kinds of grasping actions were presented, but all were without context, and what happened after the hand grasped the cup was not shown. Another type, which we called Context, showed a scene with numerous objects: a teapot, cookies, a mug, and so on. (We gave these Context clips a little bit of an Italian touch by including a jar of Nutella, the tasty Italian hazelnut-based spread.) In one Context scene, everything was neatly organized, suggesting that somebody was about to sit down to tea. In the other case, the scene was quite messy, complete with cookie crumbs and a dirty napkin, suggesting that somebody had just finished having tea. Nothing at all happens in these Context scenes: no action, only context. In the third type of clip, which we called Intention, we put together the elements of the first two types. Subjects watched a hand grasping a cup, as in the Action clip, but this time the grasping action was embedded in either the neat or the messy scene suggesting a context, as in the Context clips.

The predictions related to this experiment were relatively simple: if mirror neurons simply code the observed grasping action, activity in mirror neuron areas should be equivalent for the Action clips and for the Intention clips. If, on the other hand, mirror neurons code the intentions associated with the observed action, activity in mirror neuron areas should be greater for the Intention clips than for the Action clips, and possibly different between the two Intention clips.

The results supported the hypothesis that mirror neurons code the intentions. There was higher activity in the frontal mirror neuron area while subjects observed the grasping actions embedded in either of the two contexts, compared with the grasping action without any context. There was also higher activity while subjects observed the grasping action embedded in the context that suggested drinking, compared with the grasping action embedded in the context that suggested cleaning up. This result also made sense, since drinking is a much more primary intention than cleaning up.[17]

These results clearly support the simulation model of our ability to understand the intentions of other people. The brain cells activated while we achieve our own intentions also fire up when we distinguish between different intentions associated with the actions of other people. The form of simulation supported by mirror neurons is probably the more automatic, effortless variation of the model. Mirror neurons are located in the part of the brain that is important for motor behavior, close to the primary motor cortex sending electric signals to our muscles. This type of cell seems to have no business with a deliberate, effortful, and cognitive pretense of being in somebody else's shoes.

But *how* do mirror neurons actually predict the action that follows the observed one? How do they let us understand the intention associated with the action? My hypothesis is the following: we activate a chain of mirror neurons, such that these cells can simulate a whole sequence of simple actions—reaching for the cup, grasping it, bringing it to the mouth—that is quite simply the simulation in our brain of the intention of the human we are watching. A critical subtype of mirror neurons for this hypothesis are those cells described in the first chapter as "logically related" mirror neurons. They fire not for the same action, but for logically related ones, such as, in the monkey experiments, "grasping with the hand" and "bringing to the mouth." They are probably key neuronal elements for understanding the intentions associated with the observed action. I see you grasping a cup with a precision grip, and my precision grip mirror neurons fire. So far I am only simulating a grasping action. However, given that the context suggests drinking, the firing of other mirror neurons follows: these are my "logically related" mirror neurons that code the action of bringing the cup to the mouth. By activating this chain of mirror neurons, my brain is able to simulate the intentions of others. To put it in Gallese's words, "it is as if the other becomes another self." Or in Merleau-Ponty's words, "it is as if the other's intention inhabited my body, and mine his."[18] Mirror neurons help us reenact in our brain the intentions of other people, giving us a profound understanding of their mental states.

Can these brain cells also help us communicate with other people by facilitating the recognition and understanding of the gestures we use when we talk? Is it possible that mirror neurons may have played an even larger role in communication, as the evolutionary precursors of the neural systems that allow us to communicate with language? The answer is yes, and we'll explore how in the next chapter.

THREE

Grasping Language

If language was given to men to conceal their thoughts,
then gesture's purpose was to disclose them.
—JOHN NAPIER[1]

DO YOU SEE WHAT I AM SAYING?

I am watching my daughter talking on the telephone with one of her friends. Her arms and hands are very active—the spontaneous movements that all of us make when we talk. We even have a special word for them: "gestures." But why does my daughter gesture when talking on the telephone? After all, her friend cannot see her. My daughter is not alone: even though we know the gestures can't be seen, we all tend to gesture when we speak over the phone. Indeed, we gesture when we talk to the blind, and the congenitally blind also gesture when they talk, even though they have *never seen* other people gesturing.

Bizarre? Not really. In his book *Hand and Mind*, David McNeill argues that "gestures and language are one system," that "gestures are an integral part of language as much as are words, phrases, and sentences."[2] Note that McNeill is referring to the spontaneous arm and hand movements that are unique to the individual speaker—gestures—not to the fixed type of hand signs such as the okay sign, which we call emblems. When we cannot find the proper word to express ourselves, hand gestures can help in the retrieval of the missing word. At other times, gestures provide information that the words themselves do not provide. For instance, kids often use a dual format to explain the math concepts they are learning. One problem-solving procedure is stated with words, a different procedure with gestures. In fact, these speech-gesture mismatches indicate a transitional, expected phase in the learning process. Say a child is asked to solve the problem $5 + 4 + 3 = \_ + 3$. Her incorrect verbal response ("I added the 5, the 4, the 3, and the 3, and I got 15") may not reveal any awareness of the concept of an equation. However, if her hand moves under the left side of the equation, then stops, then moves again under the right side of the equation, the movement reveals that her mind is starting to grasp the concept that an equation has two sides that are separate but somehow related. Another example is what we call conservation tasks, which were developed by Jean Piaget to explore how children develop certain concepts. In the liquid-quantity task, the experimenter pours water from a glass to a dish. Typically, the glass is tall and relatively skinny, the dish short and wide. The child is asked whether the dish contains the same amount of water as the glass and to explain her answer. When the child answers, incorrectly, that there is less water in the dish because the level of the water in the glass was higher, her hand may form a narrow C to indicate the skinny glass and a wider C to indicate the wider dish. While her words focus only on the difference in height between glass and dish, her hands emphasize the compensatory greater width of the dish, compared with the glass. With her hands, she's catching on, and the words will soon follow.

Speech-gesture mismatches seem to indicate rich mental activity that favors the grasping of new concepts in young learners.[3] Much research has confirmed that this is the case. Typically, though not always, gestures are "ahead" of speech in these childhood mismatches. As in the equation illustration above, the gestures tend to convey the more advanced concepts. They facilitate learning. (During counting tasks, children are helped by pointing gestures, especially if they perform the gestures themselves.) Indeed, the mismatchers show a better ability to generalize recently acquired knowledge and concepts than the "skippers"—that is, kids who proceed from incorrect explanations matched in speech and gestures directly to correct explanations matched in speech and gestures.

Children are also very sensitive to the gestures of their teachers. For math problems, children are more likely to correctly repeat a procedure when the teacher's speech is matched with an appropriate gesture, compared with no gesture at all. Thus, gestures accompanying speech have a dual role of helping the speakers to express their thoughts and helping the listeners/viewers understand what is being said. It follows that mismatching gestures by the teacher get in the way of learning. Indeed, children are less likely to correctly repeat a procedure when the teacher's speech is accompanied by a mismatching gesture, compared with no gesture at all. Consider the teacher who explains the concept of the equation while pointing to each number on both sides of the equation with a series of manual gestures such as children typically use when solving a simple addition problem. This mistake only encourages the students to make the mistake made by the girl in the illustration above—adding up all of the numbers on both sides of the equation. Instead, the teacher's gestures should visually depict the two sides of the equation—perhaps a bracketing gesture with the left hand for the left side, the same gesture with the right hand for the right side. It could make a difference when teaching young children (though they will eventually catch on anyway).

As adults, our gestures are unique to each of us, but most nevertheless divide into two categories, "iconic" and "beat." Iconic gestures reflect the content of the speech they accompany. When someone verbally describes pouring wine while his hand appears to grip something and his arm raises and rotates by approximately ninety degrees at the elbow—that's an iconic gesture. Beat gestures, on the other hand, do not specifically and visually reflect what is being said. They are rhythmic hand movements that almost appear to beat out musical time with the pulsation of speech. It would seem to follow that we are less likely to employ iconic gestures when speaking on the phone or in some other situation in which the listener cannot see us, whereas the use of beat gestures would be relatively unaffected. Martha Alibali and her colleagues investigated this question with a simple experimental trick. They looked at the spontaneous gestures of speakers while they were telling the story of a cartoon to a listener. In one setup, a screen separated speaker from listener; in the other, the two were face-to-face. The results confirmed the hypothesis: the presence of the screen affected only iconic gestures, which the speakers produced at a much lower rate because they knew the gestures would not be seen. The rate of beat gestures, in contrast, was completely unaffected by the presence or absence of the screen.[4]

Another way to put the point: beat gestures seem to be more useful to the speaker, while iconic gestures are mostly produced to benefit the listener/viewer. If this thinking is correct, we can make a simple prediction with regard to brain activity, especially mirror neuron activity. The hypothesis that mirror neurons facilitate communication predicts that they would be activated more strongly during the observation of iconic gestures, which facilitate communication and understanding, than during observation of beat gestures, which are less useful for the observer. Indeed, this is exactly what we found in an fMRI experiment led by Istvan Molnar-Szakacs, at the time a graduate student working in my lab. The experiment was performed by asking a subject to watch a series of cartoons, then to narrate the cartoons while we videotaped her. We showed this videotape to subjects in the MRI scanner and found that brain areas with mirror neurons (as shown in figure 1, although only the ones in the left hemisphere) were activated when the narrator made iconic gestures, whereas a separate set of areas in a different region not known for harboring mirror neurons were activated when the narrator made beat gestures.[5]

The selective activation of mirror neurons for iconic gestures tells us that they are concerned with gestures that are important for face-to-face interactions. This point is relevant to the very contentious issue of their role in the origin of language. Let's now see why.

HAND TO MOUTH

In 1866, the Société de Linguistique de Paris banned all speculation on the origins of language. Around the same time, the British Academy warned its members not to discuss the topic, which had apparently become so contentious and speculative that the only result was endless discussion of unprovable theories. Obviously, the bans did not work. Speculation on the origins of language did not stop and probably never will, especially after the discovery of mirror neurons.

A long tradition proposes that language origins are manual and gestural. This thinking was dominant during the eighteenth century—the Enlightenment. Mirror neurons clearly support this hypothesis, for two main reasons. First is the anatomical analogy between area F5 of the macaque brain, where mirror neurons were discovered, and Broca's area, a major language center in the human brain.[6] The second reason why mirror neurons support the theory of a manual, gestural origin of language is the fact that they make the manual gestures of other people easily understandable to observers, thus providing a compelling form of communication at the gestural level.

Even before mirror neurons were discovered, scientists supporting a gestural origin of language had emphasized the strong connections between hand and mouth early in life. Why would that matter at all? The answer is expressed by a famous dictum in science: ontogeny recapitulates phylogeny. The very simple idea behind this sentence is that the embryonic and early development of a member of a species today may give us a glimpse of what happened millions of years ago during the evolution of that species. Specifically, early human development shows strong ties between the hand and the mouth. For instance, a newborn opens her mouth if one applies pressure on the palm of her hand. This Babkin reflex, as it is called, suggests that these two body parts belong to a common functional system. Furthermore, all parents know well that newborns frequently bring the hand to the mouth, introducing the fingers for sucking, and for prolonged periods of time. What may go unnoticed by parents is that newborns open the mouth *before* the hand reaches it, an act of anticipation demonstrating that the hand-to-mouth behavior is goal-oriented. Babies between nine and fifteen weeks old show systematic relations between hand and mouth movements. For instance, extension of the index finger usually co-occurs with opening of the mouth and even vocalization. Later in development, other forms of hand-to-mouth behavior are frequent. Between twenty-six and twenty-eight weeks of age there is a notable increase in such rhythmic movements of the arms and hands as banging, swaying, and waving. At the same time, babies start babbling, producing strings of the same syllables like "bababa" or "gagaga." And of course, when the baby learns to grasp with the hand, the grasped object is invariably delivered straight to the mouth.

Are the hand and the mouth coupled "equally" early in life, or is there some evidence suggesting that one is the leader and the other the follower in development (and consequently in evolution, since ontogeny recapitulates phylogeny)? Well, we have already seen that in children showing speech-gesture mismatches, the gestures generally reveal more advanced concepts than the speech. Much earlier in development, 75 percent of all babbling co-occurs with rhythmic manual activity, while approximately 40 percent of rhythmic manual activity co-occurs with babbling. These numbers suggest an earlier independence of the hand compared with the mouth. Most important, babies use communicative gestures earlier than their first words. These precocious gestures are pointing gestures and some iconic gestures, such as flapping the hands to represent birds. Given the links between mirror neurons and iconic gestures previously discussed, the use of iconic gestures very early on in development reinforces the hypothesis that mirror neurons are critical brain cells for language development and language evolution.

Children produce gesture-speech combinations—such as the word "give" coupled with pointing at an apple—earlier than word-word combinations such as "give apple." Gestures lead; speech follows. Indeed, the emergence of gesture-speech combinations typically predicts when the child will be able to use word-word combinations. Longitudinal studies on "late talkers" also suggest that gestures lead and speech follows. Some of these children catch up (the late bloomers), and others do not (the truly delayed children). What predicts the child's future in this regard is the amount of communicative gestures used. Late bloomers produce many more such gestures than truly delayed children.[7]

Taken together, all these data show that gestures precede speech and that mirror neurons are probably the critical brain cells in language development and language evolution. However, one of the defining features of human language is its syntax, which defines some sort of hierarchical structure for the words that compose a sentence. So far, we have discussed the role of mirror neurons in imitation of relatively simple actions and in coding for "intention," but what about their role in coding the hierarchical structure of actions? Any evidence in this regard would suggest that mirror neurons are also involved in more complex aspects of human language.

The developmental psychologist Patricia Greenfield has studied motor and language abilities in developing children. Patricia observed a parallel progression in the increased use of hierarchical structures in both manual activities (directed at toys and tools) and verbal communication. Drawing also on her work with chimpanzees and other great apes, she proposed—before mirror neurons were discovered!—that Broca's area was critical in the evolution and development both of manual activities and of linguistic communication. Patricia is a professor in the department of psychology at UCLA, so it was virtually inevitable that we ended up collaborating in an imaging experiment that tested the role of human mirror neuron areas in coding action of increased hierarchical complexity.

In our experiment, subjects watched an experimenter manipulating cups and rings. In some cases the sequence of actions mimicked the increasingly complex hierarchical structures displayed in spontaneous play by children; for instance, cups of different sizes may be placed one into the other, following the size order. In other cases, the manipulative actions have no obvious structure. If mirror neurons respond only to the manipulation of the objects, there should be no difference in their activity when subjects watch purposeful manipulative actions with and without a hierarchical structure. On the other hand, if mirror neurons also code the hierarchical structure of the observed action, they should respond more strongly when subjects watch actions with hierarchical structure. In our study, led again by Istvan Molnar-Szakacs, we found the stronger activity when subjects watched the manipulation with a hierarchical structure.[8] This was important not just because it validated Patricia Greenfield's theory but, above all, because it shows that mirror neurons respond to the hierarchical organization of the actions of other people. If mirror neurons can code the hierarchy of manual activities, they may also code hierarchy in other domains—for instance, in linguistic material. As we will see later in this chapter, when we humans are engaged in conversation, we tend to imitate each other's syntactical structures. In light of our brain imaging experiments on both imitation and hierarchy of action, it makes sense to assume that mirror neurons are the brain cells that help us with this imitation.

FROM BRAIN MAPPING TO BRAIN ZAPPING

Recall from chapter 2, if you will, the imaging experiments demonstrating that Broca's area was activated during both the imitation and the observation of action. Such data have been considered important evidence linking mirror neurons and language, but the activation of Broca's area in a task that does not explicitly involve language is also a double-edged sword. Is this activity simply the effect of "internal speech"? Some scientists have raised this concern. In fact, this is a classic problem with these imaging techniques: brain imaging is fascinating, but it can give us only *correlative* information. Subjects perform certain tasks while we measure how their brain activity changes during the performance. However, we have no information about the *causal* role of the observed changes in brain activity. Let me give you an example. Suppose you are in a brain imaging scanner and I ask you to wiggle the fingers of your right hand sequentially, from thumb to pinkie. You comply, activating your motor cortex, and my scanner detects this activation. However, you may entertain yourself while you do this simple motor task by silently naming the fingers you are wiggling. By doing so, you also activate your language centers, and my scanner detects this activation as well. If I have no previous knowledge about brain areas and their specialties, I must conclude that two areas in your brain are important to wiggle the fingers of the right hand, whereas in fact only one is truly essential to perform the task.

It did not seem likely that "internal speech" had produced in Broca's area exactly the pattern of activity predicted for mirror neurons—some activation for action observation, greater activation for motor execution, the highest activation for imitation—but we decided to find out for certain with transcranial magnetic stimulation (TMS). As we have seen in chapter 2, TMS works by creating a transient magnetic field under the copper coil placed on the head of the subject. This magnetic field induces an electric current in the brain region under the coil, which we call the TMS pulse. With a rapid series of these TMS pulses, the activity of that brain region is transiently disrupted—in effect, brain zapping! It may sound dangerous, but it's not. This is a powerful tool for investigating causal links between a brain area and a given function. Indeed, in healthy subjects, TMS over Broca's area induces the transient inability to speak. In our experiment, we predicted that TMS over Broca's area would impair imitation, demonstrating the causal link between this brain area and the ability to imitate.

To perform this experiment accurately and to be sure we were stimulating Broca's area, we used a technique called image-guided TMS, which allows us to see *exactly* which brain region we are stimulating without removing the skull of the subject. The way it works is as follows: The subject is first studied with an MRI. These images of the subject's brain are then transferred into the TMS lab and loaded into a system called frameless stereotaxy, which uses an infrared camera to read objects marked with a special paint. In this experiment, these "objects" are placed onto certain anatomical features of the subject's head, typically the left and right ear, the tip of the nose, and the bridge of the nose. The infrared camera reads the location in three-dimensional space of these anatomical points, and special software aligns them with the MRI images. At this point, the real anatomy of the subject is aligned with the virtual anatomy of the MRI images—very sophisticated stuff, without a doubt, and about as high-tech as neuroscience gets today.

With the stereotaxy system in place, we can move the magnetic coil over the skull of the subject and see the brain areas underneath simply by looking at the computer monitor. When we zapped Broca's area with TMS, our subjects could not imitate finger movements well. When we zapped a different brain region, the subjects imitated perfectly well. We were very excited with these results, but we had to perform a control experiment to make sure that the imitation deficits seen while zapping Broca's area were truly specific to imitation rather than to nonspecific motor deficits. To this end, we asked our subjects to perform a motor task involving finger movements exactly as in the imitation tasks, but not requiring imitation. Again zapping Broca's area, we found motor deficits only during the imitation task. This TMS experiment demonstrated a specific imitation deficit induced by a transient lesion in Broca's area, confirming that the region is essential not only for language but also for imitation.[9]

The fact that the major language area of the human brain is also a critical area for imitation *and contains mirror neurons* offers a new view of language and cognition in general. By the 1940s, say, cognitive science had become dominated by the idea that the operations of the human mind that generate language and higher cognitive functions are akin to the operations of a computer, manipulating abstract symbols on the basis of specific rules and computations. According to this view, mental operations are largely detached from the workings of the body, with the body a mere output device for commands generated by the manipulation of abstract symbols in the mind. That idea—the human mind as something quite like a computer—held sway for about half a century. Now a different view has become more and more popular. According to this alternative, our mental processes are shaped by our bodies and by the types of perceptual and motor experiences that are the product of their movement through and interaction with the surrounding world. This view is generally called embodied cognition, and the version of this theory especially dedicated to language is known as embodied semantics. The discovery of mirror neurons has strongly reinforced this hypothesis that cognition and language are embodied.

BODY HEAT

The main idea of embodied semantics is that linguistic concepts are built "bottom up" by using the sensory-motor representations necessary to enact those concepts. Let me give you an example. When we talk, we often use expressions involving actions and body parts: the *kiss* of death, *kicking* off the year, *grasping* a concept, can you give me a *hand* with this, that cost an *arm* and a *leg*, and probably hundreds more. According to the embodied semantic hypothesis, when we say, hear, or read these expressions, we actually activate the motor areas of our brain concerned with the actions performed with those body parts. When you read or say "the kiss of death," your brain activates the motor cells you activate when you actually kiss someone. (However, let's hope you won't think about death the next time you kiss someone.) There is convincing empirical evidence in line with the predictions of embodied semantics, although it generally does not involve kissing. For instance, when subjects read a sentence that implies an action moving "away" from the body, such as "close the drawer," movements of the arms *toward* the body are slowed down.

These kinds of interactions between bodily movements and linguistic material have been investigated in detail by Art Glenberg and his colleagues at the University of Wisconsin at Madison.[10] Their studies suggest that concepts are strongly tied with the biomechanical properties of bodies. Indeed, this seems to be true even when highly educated individuals are discussing extremely abstract concepts, as revealed by Eleanor Ochs and her colleagues at UCLA in a study of the scientific discussion between physicists investigating high-energy physics. Ochs and colleagues clearly show that even scientists who are trying to understand a new hypothesis ground abstract phenomena in bodily expressions. For instance, while trying to explain transitions in magnetic states owing to temperature changes in a substance, the head of the lab used downward hand gestures while saying, "When I come down in temperature, I'm in the domain state."[11] The physicist identified himself with the substance under discussion and used his arm to describe temperature changes.

Do mirror neurons participate in grounding our understanding of linguistic material in our bodies and our own actions? Vittorio Gallese and the cognitive scientist George Lakoff were the first ones to propose this hypothesis in their paper "The Brain's Concepts."[12] Lisa Aziz-Zadeh, who used to be one of my graduate students at UCLA and is now a faculty member at the University of Southern California, also in Los Angeles, performed a brain imaging experiment in my lab addressing this hypothesis specifically. Lisa asked subjects to read sentences describing hand and mouth actions—for instance, "grasp the banana" and "bite the peach"—while she measured their brain activity. Later, she showed video clips of actions performed with the hand (grasping an orange) and with the mouth (biting an apple). While subjects read the sentences and watched the actions, they activated specific areas of their brains that are known to control, respectively, the movements of the hand and the movements of the mouth. Clearly, these areas were human mirror neuron areas for hand movements and for mouth movements that were also selectively activated while subjects were reading sentences describing hand actions and mouth actions.[13] It is as if mirror neurons help us understand what we read by internally simulating the action we just read in the sentence. Lisa's experiment suggests that when we read a novel, our mirror neurons simulate the actions described in the novel, as if we were doing those actions ourselves. In its January 2007 issue, *Discover* magazine selected her study among the top six mind and brain science stories of 2006. In that issue, Lisa describes our language faculty as "intrinsically tied to the flesh."

If this is so, the role of mirror neurons in language is to transform our bodily actions from a private experience to a social experience to be shared with our fellow humans through language. Theories of both language evolution and language acquisition have always focused on some sort of iterated transmission of language or of its evolutionary precursors. The prevailing paradigm in language acquisition is that children learn from parents and teachers and will, in turn, eventually teach their own children. This transmission of knowledge is unidirectional. Similarly, the prevailing paradigm in language evolution is the one borrowed from genetics, in which the genetic endowment of a generation does not influence the endowment of the previous generation. Once again, unidirectional flow of information. Now, however, the role of mirror neurons in language invites us to look at language and its emergence with different eyes. We should look at the *coordinated activity* of interacting individuals—a *bidirectional* flow of information—to better understand the nature and emergence of human language.

CHAT ROOMS

Think about yourself engaged in a conversation, any kind of conversation. Now think about yourself engaged in a monologue, first giving a speech, then listening to a speech. Now compare these situations. Which one is "easier"? Which one comes more naturally to you? For most people, giving a speech is challenging. Most also find it demanding to follow a speech by somebody else; we have to muster all our resources of attention. In contrast, almost everyone feels quite comfortable talking with others. Even those people who find conversation and social interaction a bit of a challenge find monologues more challenging than conversation. Why should this be? From the standpoint of the *cognitive* demands of monologues and conversations, it makes no sense that the complicated give-and-take of a conversation should be easier than delivering a monologue. If anything, it should be exactly the opposite.

Consider the issue of planning what to say. You can plan a speech from start to end, but you cannot plan a conversation. Who knows what the other person is going to say? Our powers of discerning intention, even with our mirror neurons, are not that omniscient. This difference alone should make giving a speech much easier than conversation. A related issue is timing. A speaker engaged in a monologue has complete control over the pace of the speech. You can speed up or slow down, take long pauses, or do whatever is necessary to make your presentation more powerful. But people engaged in a conversation do not enjoy any such freedom. Indeed, taking turns during a conversation is a really fast business. The pause between the end of the utterance of the speaker and the beginning of the next utterance from a different speaker is about a tenth of a second. Longer pauses can feel unbearably long to people engaged in the conversation. So, again, this difference in the control of timing should make monologues easier than conversations.

That's not all. At least two other major factors should make monologues easier. The first has to do with the type of statements people make. Monologues tend to have complete and well-formed sentences, while conversational utterances are almost invariably fragments that make it necessary for the listener to guess missing information. Then there's the rapid-fire switching back and forth from speaking to listening required in conversation, a highly demanding cognitive operation.

For all these reasons, conversation should be much more difficult than delivering a monologue. However, the exact opposite is the case.[14] Conversation is easier than monologue, and I believe the explanation is rooted in mirror neurons and imitation. During conversation, we imitate each other's expressions, even each other's syntactic constructions. We also automatically and interactively negotiate the meaning of certain words, so that those words assume a very precise meaning within the context of a specific conversation rather than the meaning we would get from a dictionary. This is why overhearing a conversation does not mean that we will be able to make sense of it.[15] Try to read a transcript of a computer chat room in which you did not participate: it feels as if you have no clue as to what is being talked about—and you probably don't. You may think that imitation has no role in these virtual conversations, because people do not see each other. However, we can and do imitate words, syntactic constructions, and so on. For instance, if one person engaged in a dialogue uses the word "sofa" rather than the word "couch," the other person engaged in the dialogue will do the same.

There are other forms of imitation and interactive alignment in a face-to-face conversation. Meaning and turn-taking are automatically negotiated; simultaneous gestures, eye orientation, and body rotations are very important for helping us make sense of what is being said. These nonverbal forms of communication easily fall into patterns. Although it may feel as if we always look at our fellow interlocutors, detailed analyses of videotaped spontaneous conversations reveal that at the beginning of the speaker's turn, the listener rarely looks at the speaker's *eyes*. Soon thereafter, the listener gazes at the speaker's eyes.[16] At this precise moment of *mutual gaze*, the speaker tends to start a new sentence without completing the one in progress. It is almost as if by looking straight into the other person's eyes, the hearer is saying "Go ahead now, it's your turn to talk and I won't interrupt (for a few seconds . . .)."

Simply put, both the words and the actions in a conversation tend to be part of a coordinated, joint activity with a common goal, and this dance of dialogue is natural and easy for us. But none of it is generally studied by traditional linguistics. More to the point, such a dance is also *exactly* the kind of social interaction that mirror neurons facilitate through imitation.

Every conversation is a coordinated activity with a common goal, and all re-create to a degree the evolution of a new language. Indeed, the fact that some words in a conversation assume specific meanings determined by tacit, mutual consensus shows us how the joint power of imitation and innovation creates communication. One of the most startling examples in support of this idea is the spontaneous but fully developed sign language created by deaf children in Nicaragua schools in the late 1970s and 1980s. Before that time, the deaf in Nicaragua were largely isolated, communicating with friends and family through simple gestures and "homemade," ad hoc sign systems. Then the Sandinista revolution encouraged centers for special education of deaf children. Hundreds were enrolled in two schools in the Managua area—a critical mass, it turned out. While interacting in the school yard, on the buses, and in the streets, these children progressively developed a shared sign language by combining gestures from their individual systems. Initially, this was a relatively simple language with a simple grammar and few synonyms—what is called a pidgin language. Later on, younger kids who had been taught this simple language by the older kids developed a more sophisticated, well-defined, stable, and full-blown sign language that is now known as Idioma de Señas de Nicaragua. Ironically, the staff at the school could not understand what the children were signing to each other and had to rely on the external help of Judy Kegl, an American linguist expert in American Sign Language, to figure out what was going on.[17]

This story of spontaneous language development is famous all over the world. Some scientists have interpreted the phenomenon as evidence that humans are hardwired for language acquisition.[18] I believe that mirror neurons provide a simpler explanation, for they make it possible to automatically and deeply understand the hand movements and gestures of other people and to imitate those gestures. This is a fundamental starting point for creating a set of gestures as the basis of a relatively simple sign language. From this basis, it is also relatively simple to settle on, through the reciprocal imitation facilitated by mirror neurons, a more complex structure of gestures that build full-blown sign language. The key element that made all this possible in Nicaragua was the face-to-face interaction among kids throughout the day. This is the sort of context in which mirror neurons can work their magic to a maximum effect.

I am not alone in this hypothesis. Other scientists have emphasized the role of imitation in the emergence and acquisition of language. The psychologist Michael Tomasello has noticed that children learn the concrete linguistic expressions of their language *through imitation*, and they tend to stick with those expressions, repeating them very frequently. Some children go through a stage in which they tend to use the expression "I think" with the meaning of something like "maybe." These children virtually never use other forms of this expression, for instance "he thinks" or "I don't think" or "I thought" or even "I think that." The repeated use of a fixed expression clearly suggests that imitation, and not an instinct for grammar, helps children acquire language. Later on, they start combining these acquired fixed expressions and are able to build more sophisticated language forms.[19]

Other scientists have explored the emergence of communication with well-controlled experiments in the laboratory. A typical way of studying how humans "invent a language" is by engaging participants in a collaborative game, perhaps one in which two players are trying to communicate each other's position in a maze. In these situations, participants tend to come up with new meanings for existing words and to quickly adopt these new meanings, as if creating some sort of sublanguage through imitation. In a variation of such games, participants cannot even talk to each other, and can communicate only graphically, by drawing lines. Sometimes the scratch pad provided to the subjects for this graphic conversation moves only vertically as subjects draw on it, thus forcing them to create entirely novel forms of visual communication. Even in these cases, participants are able to communicate by coordinating each other's activity through reciprocal imitation.[20]

This discussion leads to an obvious question: If imitation is such a key factor in language acquisition and even in the emergence of language, do the neural mechanisms of mirroring occur not only for actions (which we know they do, including with sign languages) but also for speech *sounds*? After all, the way children learn language is initially and mainly based on the spoken form of language. The final section of this chapter deals with this question.

MIRRORING SPEECH AND OTHER SOUNDS

When monkeys hear certain sounds that are typically associated with certain actions, for instance breaking a peanut, mirror neurons fire. This we learned in the first chapter. Is there any evidence that human mirror neurons do the same? While still a graduate student in my lab, Lisa Aziz-Zadeh used TMS to measure the excitability of motor cells in the human brain while inactive subjects were listening to different sounds. As predicted, she found that subjects had higher motor excitability when listening to action sounds, such as tearing paper and typing on a keyboard, compared with the response to other kinds of sounds—for instance, thunder. Moreover, the higher excitability was restricted to those muscles involved in the actions producing the sounds. When the subject was listening to the sound of tearing paper, hand muscles were more excitable than foot muscles. This was the same phenomenon of "motor resonance" Luciano Fadiga demonstrated for observed actions in the experiment described in chapter 2. Also predictably, a brain imaging study led by Christian Keysers in Holland demonstrated activation of mirror neuron areas while subjects listened to action sounds.[21]

Although these experiments show nicely a link between human mirror neurons and sounds, they do not tell us whether the perception of *speech* sounds, specifically, is facilitated by some form of neural mirroring. However, a well-known behavioral effect on speech perception called the McGurk effect suggests some form of mirroring. When subjects are listening to discrete syllables played by speakerphones—for instance, *ba*—while they are also watching the videotape of somebody moving the lips as if to say *ga*, they perceive the sound of neither *ba* nor *ga*. Instead, they hear a third syllable, *da*, that has not been presented to them at all.[22] The McGurk effect shows that watching the moving lips of a speaker evokes in our brain the speech sounds that would be spoken by that person; if we are simultaneously listening to a different speech sound, the two sounds merge in the brain and form a third sound that we have not heard at all.

Several years ago at the Haskins Laboratory at Yale, Alvin Liberman and his colleagues were trying to build devices that would transform text into spoken words so that war veterans who had lost their eyesight would be able to "read" books and newspapers. To their dismay, Liberman and colleagues found that the veterans' perception of the device's speech output was unbearably slow, much slower even than the perception of distorted human speech. This observation inspired the Yale team to propose a theory of speech perception according to which speech sounds are understood not so much as sounds, but rather as "articulatory gestures"—that is, as the intended motor plans necessary to speak.[23] This motor theory of speech perception basically suggests that the way our brain perceives other people's speech is by simulating that we are talking ourselves!

Immediately after mirror neurons were discovered in Parma, Giacomo Rizzolatti told Luciano Fadiga that the properties of those neurons reminded him of the motor theory of speech perception of Alvin Liberman. This remark inspired Fadiga to use TMS to test Liberman's motor theory. In this experiment, as the subjects listened to words through earphones, Fadiga and his colleagues stimulated the sector of the motor cortex that controls tongue muscles while recording the tongue muscle twitches induced by the stimulation. They used two main types of words. One required strong tongue movements when produced (double *r*, such as "*terra*," or "ground" in Italian). The other type required only a slight tongue movement when produced (double *f*, such as "*baffo*," or "mustache" in Italian). The motor theory of speech perception predicts that while subjects listen to words that require strong tongue movements, the stimulation over the tongue motor cortex should produce stronger muscular twitches in the tongue compared with listening to words requiring little tongue movement. The results confirmed it.[24] In a manner analogous to Fadiga's previous experiments on motor resonance for grasping actions, the experiment demonstrated that while listening to other people talk, listeners mirror the speaker with their tongues.

Following this work, Stephen Wilson, then a graduate student in my lab, used fMRI to look at brain activation while subjects said a series of syllables aloud and listened through earphones to other people speaking the same syllables. In every subject studied, the same speech motor area that was activated during speaking was also activated during listening.[25] Both Fadiga's and Wilson's studies, then, clearly show that when we listen to others, our motor speech brain areas are activated as if we are talking. But is this activation of mirror neurons for speech necessary to perceive speech? Ingo Meister, a German neurologist who spent a year in my lab studying the links between the mirror neuron system and language, performed a TMS experiment to answer this question. In the usual way, a copper coil was placed on subjects' heads in order to induce transient lesions in the motor speech area identified by Stephen Wilson. Thus "incapacitated," would the subjects nevertheless be able to *understand* speech? Ingo and the rest of us believed the answer would be no, and this was indeed the case. When the TMS pulses knocked down the subjects' motor speech areas, their ability to perceive speech sounds was also reduced.[26] The mirroring of other people's speech is actually necessary for us to perceive it.

These studies have opened a whole new field of experiments on neural mirroring of vocalization. In one recent experiment, scientists found that listening to vocalizations of amusement and triumph—laughter and excited shouts, say—activates the same motor areas required for smiling.[27] This result suggests mirroring for the sharing of positive emotions expressed through vocalization. Such a mirroring mechanism would seem to be essential for establishing cohesive bonds within social groups. The next question for investigation follows automatically: What is the role of mirror neurons in the various forms of empathy that characterize human behavior?

FOUR

See Me, Feel Me

*When we see a stroke aimed, and just ready to fall upon
the leg or arm of another person, we naturally shrink
and draw back our leg or our own arm; and when it does
fall, we feel it in some measure, and are hurt by it as
well as the sufferer.* —ADAM SMITH (1759)[1]

ZIDANE'S HEAD-BUTT

I am in Italy. It is early August 2006, and a month ago Italy won the World Cup of soccer by defeating France on penalty kicks following a 1–1 tie. A key episode in the Italian triumph was the sudden ejection of Zinédine Zidane, France's world-class player, just a few minutes before the end of the extra time that preceded the penalty kicks. The cause was his savage head-butting of Marco Materazzi, an Italian player, in the chest—an act of folly seen live by more than a billion people. Zidane was the designated French player for penalty kicks and had already scored the penalty kick that gave

France a temporary lead in the game. In the end, Italy won the penalty kicks, with one penalty kick missed by a French player. It is widely thought that Zidane's premature exit from the game was a decisive factor in the outcome.

Now, a month later, we are enjoying a typical Italian summertime dinner party, with lots of relatives gathered at my uncle's beach house at San Felice Circeo, a town about sixty miles south of Rome. (I've been living in Los Angeles for many years, and every time I return to Italy for a short visit, we have a good reason for a family feast.) Just before dinner, my cousin-in-law is flipping TV channels and finds a broadcast of the World Cup final. He informs me that one channel or another shows the whole game at least once a week, and I am not surprised. The last time Italy won the World Cup was twenty-four years ago. Italians want to relive every moment of their triumph, because who knows when we will have another one. Watching the game again, I know exactly what is going to happen. Still, I find myself experiencing strong emotions when Zidane head-butts Materazzi. I wince at Materazzi's pain. Read the quote from Adam Smith at the beginning of this chapter again: well over two hundred years ago, he described the phenomenon nicely. I also feel enraged all over again at Zidane for his act of aggression. A few minutes later I watch the French player Trézéguet missing his penalty kick. The ball hits the post and bounces away—the mistake that gave the Italian squad the World Cup.

And now my point for this book: watching the Zidane head-butt, I feel my original emotions of the moment almost as powerfully as ever, but I feel no emotions watching the missed penalty kick, which, in the long run, one could argue was far more important than Zidane's head-butt. Why did only Zidane's head-butt trigger my strong emotional reaction a month later? When I see the head-butt, I am watching two bodies colliding, a head hitting a chest, and the faces of the two men roiled by strong emotion. I have an immediate, unmediated, and automatic understanding of what these two individuals feel. When I see the ball hitting the post and bouncing away, however, I am watching two inanimate objects. In theory, the missed kick is more important, but my brain isn't dealing in theories. My brain is dealing with what it *sees*, and what it sees determines what I *feel*.

I believe the most likely explanation of my empathy for the emotions of the head-butting episode is some neural mechanism for mirroring in my brain. My friends in Giacomo Rizzolatti's lab in Parma agree; among them, Vittorio Gallese was the first to propose a role of mirror neurons in both understanding and empathizing with the emotions of other people. Gallese, the investigator whose interest in philosophy introduced the team to the important work of the phenomenologist Maurice Merleau-Ponty, also pointed out the groundbreaking work on empathy by the German psychologist Theodor Lipps at the beginning of the twentieth century, work that, in retrospect, points directly at a role for mirror neurons. "Empathy" is actually a later English translation of the German word "*Einfühlung*," which Lipps initially proposed to describe the relationship between a work of art and its observer. He subsequently extended this concept to interactions between people: he construed our perception of the movements of others as a form of inner imitation and provided the example of watching an acrobat suspended on a wire high above the seats at the circus. When we watch the acrobat on the wire, Lipps says, we feel ourselves inside the acrobat. His phenomenological description of watching the acrobat is eerily predictive of the pattern of activity displayed by mirror neurons that fire both when we grasp and when we see someone else grasping, as if we were inside that person.[2]

Empathy plays a fundamental role in our social lives. It allows us to share emotions, experiences, needs, and goals. Not surprisingly, there is much empirical evidence suggesting a strong link between mirror neurons (or some general forms of neural mirroring) and empathy. This evidence has been gathered using different methodologies in neuroscience, from brain imaging to the study of brain-damaged patients, even looking at the data from depth electrodes implanted in neurosurgical patients. Before digging into the details of this work, though, I need to present the series of careful studies of human behavior by social psychologists. This work was the very first evidence suggesting links between mirroring and empathy.

HUMANS OR CHAMELEONS?

Sometimes I think of humans as chameleons, and I am not the first to have made the comparison. We have an instinct to imitate one another—to synchronize our bodies, our actions, even the way we speak to each other—just as we have seen in the previous chapters. This phenomenon has been studied in a host of ways, from clever experiments to detailed observations of human behavior. As Elaine Hatfield, John Cacioppo, and Richard L. Rapson put it in their wonderful book, *Emotional Contagion*, "people imitate others' expression of pain, laughter, smiling, affection, embarrassment, discomfort, disgust, stuttering, reaching with effort, and the like, in a broad range of situations. Such mimicry . . . is a communicative act, conveying a rapid and precise nonverbal message to another person."[3]

This rapid and precise nonverbal synchrony we "enjoy" with our fellows often has an emotional component, as in Frank Bernieri's videotapes of young couples teaching each other some made-up words. He found that the couples who demonstrated the greatest motor synchrony also had the strongest emotional rapport with each other. A study on the role of an interviewer's warmth on an interviewee's reaction shows that warm interviewers—those instructed to lean forward, smile, and nod—elicited leaning-forward movements, smiles, and nods in the interviewees. Such motor mimicry seems to have not only a communicative role, but even a perceptual one, as demonstrated by Ulf Dimberg's measurements of the activity of facial muscles of subjects who were looking at happy or angry faces. The activity of cheek muscles that we contract when smiling increased when these subjects were watching happy faces, and the activity of brow muscles that we contract when angry increased while they were watching angry faces.[4] Remember, there was no face-to-face interaction in this experiment—the subjects were looking at pictures. What, then, was the role of mimicry in that situation? The answer comes from a study led by Paula Niedenthal, an American social psychologist who lives and works in France. In her experiment, two groups of participants were asked to detect changes in the facial expressions of other people. The key was that one group was prevented from freely moving their own faces because they were holding a pencil between the teeth. Try it. The pencil severely restricts the ability to smile, frown, and do everything else we do with our faces throughout the day. Moreover, the pencil severely restricts mimicry. Thus it is no surprise to learn that the participants holding the pencils between their teeth were much less effi-cient in detecting changes in emotional facial expressions than were participants who were free to mimic the observed expressions.[5] Mimicking others is not just a form of communi-cating nonverbally; it helps us to perceive others' expressions (and therefore their emotions) in the first place.

This seems counterintuitive. Wouldn't we expect that in order to imitate an observed facial emotional expression, we first should be able to *recognize* it? Only if we assume that deliberate, explicit recognition must necessarily precede mim-icry. Our only proof of this is an age-old theory, and the exis-tence of mirror neurons provides an alternative explanation: the idea that the mimicry actually *precedes* and helps the recognition. This is what I believe happens: mirror neurons provide an unreflective, automatic simulation (or "inner imitation," as I have sometimes phrased it here) of the facial expressions of other people, and this process of simulation does not require explicit, deliberate recognition of the expression mimicked. Simultaneously, mirror neurons send signals to the emotional centers located in the limbic system of the brain. The neural activity in the limbic system triggered by these signals from mirror neurons allows us to feel the emotions associated with the observed facial expressions—the happiness associated with a smile, the sadness associated with a frown. Only *after* we feel these emotions internally are we able to explicitly recognize them. When a participant is asked to hold a pencil between his teeth, the motor activity required by this action interferes with the motor activity of mirror neurons that would mimic the observed facial expressions. The subsequent cascade of neural activations that would lead to explicit recognition of emotions is also disrupted.

If this account of how mimicry can support recognition of emotions is correct, it follows that good imitators should also be good at recognizing emotions, and so endowed with a greater empathy for others. And if *this* is true, we should observe a relationship between the tendency to imitate others and the ability to empathize with them. As it happens, this is exactly the hypothesis tested by the social psychologists Tanya Chartrand and John Bargh.[6] In their first experiment, a subject was asked to choose pictures from a set of photographs. Sitting in the same room was an experimenter *pretending* to be another subject. (In experimental jargon, this was the "confederate.") The cover story was that the researchers needed some pictures for a psychological test and wanted to know which pictures the subject considered more stimulating. In fact, while the real subject was choosing the pictures, the confederate was engaged in a very deliberate action, either rubbing his face or shaking his foot. Subjects were videotaped, and their motor behavior was measured. Analyzing the videotapes, Chartrand and Bargh found that the subjects unconsciously mimicked the action of the confederate also in the room. Subjects sharing the room with the face-rubbing confederate rubbed their own faces more than subjects who shared the room with foot-shaking confederates, and vice versa.

In a second experiment, Chartrand and Bargh tested the hypothesis that one of the functions of the "chameleon effect" is to increase the likelihood that two individuals will readily like each other. Again, participants were asked to choose pictures in the company of a confederate pretending to be another participant. This time the cover task required participant and confederate to take turns in describing what they saw in various photos. All the while, the confederate either imitated the spontaneous postures, movements, and mannerisms of the subject or kept a neutral posture. At the end of these interactions, the participants were asked to complete a questionnaire to report how much they liked the other participant (that is, the confederate) and how smoothly they thought the interaction had gone. You can predict the results by now: the participants who were mimicked by the confederates liked those confederates much more than the participants who were not imitated. Further, the mimicked subjects rated the smoothness of the interaction higher than the participants who were not imitated. The experiment clearly shows that imitation and "liking" tend to go together. When someone is imitating us, we tend to like that person more. Is this *why* we have such an automatic tendency to mimic one another? I think so.

In their final, most critical experiment, Chartrand and Bargh tested the hypothesis that the more you are a chameleon, the more you are concerned with the feelings of other people—that is, the more empathy you have. The setting of this third experiment was the same as in the first setup, with the confederate either rubbing the face or shaking the foot. The novel aspect of this last experiment was that the participants responded to a questionnaire that measured their empathic tendencies. Now Chartrand and Bargh found a strong correlation between the degree of imitative behavior displayed by the participants and their tendency to empathize. The more the subject imitated the face rubbing or the foot shaking, the more that subject was an empathic individual. This result suggests that, through imitation and mimicry, we are able to feel what other people feel. By being able to feel what others feel, we are also able to respond compassionately to their emotional states.

These are well-designed, convincing studies, and there are many others. In fact, a thorough discussion of all the behavioral data suggesting the strong link between imitation and empathy would require a book of its own, but I do want to

mention two findings that represent the far ends of a continuum. On the one hand, we know that couples have a "higher
facial similarity" (they look more like each other) after a
quarter century of married life than at the time of their mar-
riage. This effect also correlates with the quality of the mar-
riage: the higher the quality, the higher the facial similarity.
This is no surprise, really. Loving, sharing, and living together
make us look more and more similar to each other. The
spouse becomes a second self. On the other hand, there are
the "consequences of facial loss," according to Jonathan Cole,
a British neuropsychologist who investigates the subjective effects of facial differences. Among his patients, those born
with Moebius syndrome, a congenital inability to move the
muscles of the face, report not only the altered ability to communicate felt emotions, but also the inability to read emotions in others. One patient says, "For the face of the other
requires me to respond and enter into a relationship, but a relationship I cannot fully control." Our esteemed friend Maurice Merleau-Ponty wrote, "I live in the facial expression of
the other, as I feel him living in mine." Alas, Cole's patients
cannot live in the facial expression of the other, because they
are unable to move their own facial muscles. This inability,
and the consequent inability to mirror others' facial expressions, breaks down any form of emotional interaction and
makes impossible a deeply felt understanding of others'
emotions.[7]

So the case for a link between the neural systems for imitation (the mirror neuron system) and the neural systems for

emotions (the limbic system) is strong. However, these two systems are quite different within the brain. How do they get together? What is the neural pathway?

EMPATHIC MIRRORS

The features are given to man as the means by which he shall express his emotions. —SIR ARTHUR CONAN DOYLE

To be perfectly honest, the problem—how are brain areas containing mirror neurons connected with limbic areas concerned with emotions?—was brought to my attention in the fall of 2000 by a student in a seminar on imitation and mirror neurons. In my presentation, I had suggested a hypothetical relationship between mirror neurons and the ability to empathize. (At that time, the relationship was still quite hypothetical. We were just beginning to collect the empirical data.) In the Q and A that followed, this man asked if I knew of any anatomical connections between the mirror neuron system and the limbic system, given my hypothesis on the role of mirror neurons in our understanding of the emotions of others. My feeble answer to this brilliant question was that I basically had no idea—and definitely needed to look into it. (Incidentally, this episode is a good example of why I like to give seminars. Students—and colleagues, of course—often give me good ideas and force me to reexamine an issue I think I've already figured out.)

When we try to understand how different brain regions

"talk" to each other, a good place to start is with the anatomy of the organ. A basic fact of this anatomy is that in order to communicate with each other, brain cells must be connected in some way. In one way, of course, every cell is connected to every other cell, just as the tiniest town in this vast country—the United States—is connected to every other town by the road system. You *can* get from A to Z; the route may be circuitous, but you can do it. Likewise in the brain, but in theory only, because the maze of connections is exponentially—infinitely, it is fair to say—more complex than the U.S. or any other road system. In order for one region of the brain to talk with another, a pretty direct pathway is required—an interstate of some sort, if you will. That's what I needed to identify, and when I finally found the time to study the matter, one brain region and only one stood out with well-documented anatomical connections to both mirror neuron and limbic areas.[8] It is called the insula, the name a Latin word that means "island." The shape of the insula may remind us of an island (at least it must have reminded someone of an island, back when areas of the brain were being named), but otherwise the name is a misnomer. Functionally, the insula is not an island at all—it has a remarkable pattern of anatomical connections with a large number of other brain areas. This realization was very exciting for me. I had finally found evidence for an anatomical pathway connecting mirror neuron areas and limbic areas, a pathway that might support my hypothesis that we understand the emotions of other people thanks to our own mirror neurons, which are activated by the sight of someone else's smiling or frowning face.

After the initial excitement, however, I sobered up. The existence of anatomical connections is a nice prerequisite for two brain regions that are supposed to talk to each other. However, the anatomy does not tell us what kind of conversation is going on. What I needed was a brain imaging experiment that supported my hypothesis. Given the links between imitation and the mirror neuron system in humans discussed in chapter 2, and given the large amount of behavioral data on imitation and empathy discussed previously in this chapter, I decided to look at the brain activity of healthy volunteers while they were either watching or *imitating* pictures of faces expressing the basic emotions of fear, sadness, anger, happiness, surprise, and disgust. The idea was relatively simple: if indeed mirror neurons are communicating with the "emotional" brain areas in the limbic system through the insula, the fMRI technology should show simultaneous activation of all three areas—mirror neurons, limbic system, and insula—while the subjects are just watching the faces expressing emotion. Furthermore, if mirror neurons are sending the signals, we should find an increase in brain activity in those subjects who were also imitating the expressions. This increase should be observed not only in mirror neuron areas but also in the insula and limbic areas, because the increased activity in mirror neuron areas should spread to the others, which are receiving signals from mirror neurons. This last point is the key: the expected activity in the mirror neuron areas during imitation should spread.

That was our hypothesis, and the results confirmed my two predictions. Indeed, mirror neuron areas, the insula, and emo-

tional brain areas in the limbic system, particularly the amyg-
dala—a limbic structure highly responsive to faces—were
activated while subjects were observing the faces, and the
activity increased in those subjects who were also imitating
what they saw. These results clearly supported the idea that
mirror neuron areas help us understand the emotions of other
people by some form of inner imitation. According to this
mirror neuron hypothesis of empathy, our mirror neurons fire
when we see others expressing their emotions, as if we were
making those facial expressions ourselves. By means of this firing, the neurons also send signals to emotional brain centers
in the limbic system to make us feel what other people feel.

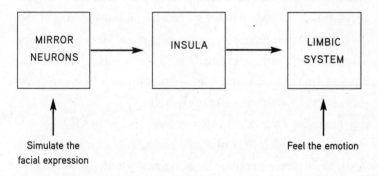

*Figure 2: Neural mechanisms for empathy. Mirror neurons provide an inner
imitation, or simulation, of the observed facial expression. They send signals
through the insula to the limbic system, which provides the feeling of the observed emotion.*

In his famous short story "The Purloined Letter," Edgar
Allan Poe writes, through the words of the protagonist C. August Dupin, "When I wish to find out how wise, or how stu-

pid, or how good, or how wicked is any one, or what are his thoughts at the moment, I fashion the expression of my face, as accurately as possible, in accordance with the expression of his, and then wait to see what thoughts or sentiments arise in my mind or heart, as if to match or correspond with the expression." What remarkable prescience! Poe could not have picked a better way to get into the inner lives of his characters. However, he was not alone. In the scientific literature on emotions, the theory that emotional experience is shaped by changes in facial musculature—the "facial feedback hypothesis"—has a long history. Charles Darwin and William James were among the first ones to write about it (although Poe predated both by several decades). Darwin writes, "The free expression by outward signs of an emotion intensifies it. On the other hand, the repression, as far as is possible, of all outward signs softens our emotions." To James, this phenomenon means that "our mental life is knit up with our corporeal frame, in the strictest sense of the term."[9]

Plenty of empirical evidence supports the facial feedback hypothesis, which in turn aligns beautifully with our investigations of mirror neurons. By firing as if we are actually making those facial expressions we are simply observing, these neurons provide the mechanism for simulated facial feedback. This simulation process is not an effortful, deliberate pretense of being in somebody else's shoes. It is an *effortless*, automatic, and unconscious inner mirroring.

We published the results of our experiment in the *Proceedings of the National Academy of Sciences*.[10] The paper attracted quite a bit of media attention, even from a couple of major TV channels. To catch the attention of readers, some newspapers and magazines borrowed a line that was famously used by former president Bill Clinton while interacting with AIDS demonstrators during his presidential campaign: "I feel your pain." (This was piling it on. Clinton's detractors, along with every comedian in the country, used the line to tease him unmercifully for years.) Of course, our experiment had not focused specifically on what happens when we see other people in pain. Following our study, however, a series of other studies did.

I FEEL YOUR PAIN

On occasion, the treatment of choice for chronic depression, obsessive-compulsive disorders, and certain other psychiatric disorders requires the removal of the cingulate cortex, a region of the neocortex closely connected with the premotor cortex. Prior to the procedure itself, and with the patient's permission, of course, the neurosurgeons may be able to use electrodes implanted in the depth of the brain as part of the neurosurgical procedure to examine activity at the level of the single cell. (For ethical reasons, the classic single-cell investigations with the macaque monkeys are off-limits—with rare exceptions. This surgery is one of them. Perhaps the most significant group of exceptions is that offered by epileptics, as we will see in chapter 7.) Obviously, the location of the electrodes in these patients is dictated exclusively by medical reasons, not by experimental curiosity. Still, with the help of the patients, it is possible to collect unique and extremely valuable information.

Many functions are associated with activity in the cingulate cortex. One is response to painful stimulation. William Hutchison and his colleagues at the University of Toronto studied some cells in the human cingulate cortex that responded selectively to painful stimulation applied to the patients—to pinpricks, in one instance. The researchers also found that one of these cells responded to the *sight* of pinpricks applied to somebody else—in this case it was the examiner's fingers.[11] This cell seems like a mirror neuron, except that—in contrast with the mirror neurons I described so far—Hutchison's cell seems specialized in processing pain. Mirror neurons typically fire for actions, not for pain. That is, they are primarily motor neurons (although they obviously have important sensory properties). Indeed—with our imaging experiment on mirroring emotions described earlier in this chapter—we suggested that we mirror the emotions of other people by activating first mirror neurons for facial expressions (thus, motor neurons), which in turn activate our emotional brain centers. According to our model, mirroring of emotion is mediated by action simulation (in our case, the facial expression, as also shown in figure 2). The human cingulate neuron for pain described by Hutchison and colleagues raised the possibility of simulation mechanisms for pain that bypass the motor behavior associated with pain.

This limited depth electrode research, however, could not

sample the whole brain. Therefore the responses in the cingulate cortex that are specific to pain do not tell us whether or not mirror neurons in motor areas are also activated. Those motor areas are definitely activated when we jerk a hand away from the hot burner on the stove, but are they also activated when we simply witness someone else brushing a hand against the hot stovetop? If our thinking about the role of mirror neurons is correct, the answer is yes. A full-simulation brain mechanism would mirror not only the pain, but also the motor reaction of the person we are watching.

To test this hypothesis, Salvatore Aglioti and his associates at the University of Rome performed an experiment using transcranial magnetic stimulation.[12] Building on the basic observation that mirror neurons are activated by the sight of others' actions, Aglioti and colleagues used the copper coil to measure the excitability of the motor cortex while participants watched videos of needles penetrating the hands and feet of other people. For purposes of comparison and control, the participants also watched harmless Q-tips gently brushed over the hands and feet, as well as needles penetrating not hands and feet, but tomatoes. All the while, the Aglioti team measured the excitability of a muscle in the hand that aids in moving the hand toward the needle. They also measured excitability in a neighboring hand muscle that has no role in moving the hand either toward or away from the needle.

The prediction: an empathic pain response on the part of the subject would produce a *reduced* excitability of the muscle that would move the hand *toward* the needle. The results: as expected. The motor cortex controlling the muscle that would move the hand toward the needle was less excitable when subjects were watching the needles penetrate hands, compared with watching them penetrate feet or tomatoes or watching the Q-tips. The decreased excitability during the observation of pain was also specific to the muscles penetrated by the needles. Neighboring muscles in the hand did not change their excitability. Furthermore, subjects were asked after the experiment to rate the intensity of the pain felt by the individuals observed in the videos. Aglioti and colleagues found that the lower the motor excitability in subjects' muscles during the experiment, the higher they rated the pain. That is, the more strongly they empathized with the observed pain, the more strongly their brain simulated a "withdrawing" action from the needle.

This experiment demonstrates that our brain produces a *full simulation*—even the motor component—of the observed painful experiences of other people. Although we commonly think of pain as a fundamentally private experience, our brain actually treats it as an experience shared with others. This neural mechanism is essential for building social ties. It is also very likely that these forms of resonance with the painful experiences of others are relatively early mechanisms of empathy, from an evolutionary and developmental point of view. More abstract forms of empathy may rely less on somatic mirroring and more on affective mirroring. In other words, in more abstract situations we may be able to empathize by mirroring the affective aspect of pain. For instance, how do we empathize with others in situations in which we do not see their facial expressions, their body postures, their gestures?

How do we empathize with the victims of terrible tragedies, such as Hurricane Katrina or the Christmas Eve tsunami in Southeast Asia? In London, brain mapper Tanya Singer addressed specifically this question by studying the empathic responses in couples.[13] In this setup, the woman was lying in the scanner while her husband, fiancé, or boyfriend sat in a chair nearby, each of them hooked to an electrode on the hand through which Singer could deliver electric shocks. A colored arrow flashing on their respective computer monitors alerted the couple ahead of time who was going to receive the upcoming shock, the man or the woman, and also indicated the severity of the shock.

When Singer looked at the brain responses of her female subjects, she found that the subject who was shocked herself showed increased activity in somatosensory areas that process tactile information (owing to the sensory stimulation of the shock applied on the hand of the subject) and in brain areas that process the emotional aspect of pain (the unpleasant feeling associated with it; among these areas was the same cingulate cortex in which Hutchison and colleagues observed the brain cell responding to the pinpricks). When the women in the scanner knew that their partners were going to be shocked, they activated *only* the affective areas relevant to pain, not the sensory regions. The key point here is that these subjects do not see the physical harm delivered to the hand of the partner. They do not see the man's face registering a painful expression. They hear no yelps of pain. Their foreknowledge is rather abstract: a colored arrow on a computer monitor is the only information they have regarding the pain felt by the partner. Yet even in this artificial situation the brains of Singer's female subjects mirrored the affective aspect of the pain experienced by the men.

It seems as if our brain is *built* for mirroring, and that only through mirroring—through the simulation in our brain of the felt experience of other minds—do we deeply understand what other people are feeling.[14]

MATERNAL EMPATHY

If mirroring is such a powerful mechanism for understanding the emotional states of other people and empathizing with them, one would expect lots of mirroring between parents and their children. There is certainly plenty of behavioral evidence supporting this supposition. I noted in chapter 2 that newborns are instinctively imitating movements from their first hours. Infants as young as ten weeks old spontaneously imitate some rudimentary features of happy and angry expressions displayed by their mothers. Nine-month-old infants comprehensively mirror facial expressions of joy and sadness. And, of course, mothers also imitate the facial expressions of their infants: from day one, an open mouth elicits an open mouth.[15] Mothers also tend to synchronize their movements more with their own children than with unrelated children.[16] In classic attachment theory, maternal sensitivity is even defined as the mother's disposition to answer in a contingent way to her child's needs. Mirroring allows her to achieve a powerful affective attunement, and the maternal capacity to mirror the infant's internal states probably takes many forms.

The role of mirror neurons in maternal empathy is still largely unknown, although it is probable that these cells are important for this crucial function. My group at UCLA is literally taking the first steps in the attempt to understand the neurobiological mechanism underlying maternal empathy. In a collaborative project with a group of Italian neuroscientists and psychologists in Rome, I have recently studied the neural responses of moms looking at pictures and imitating the expressions of their own and another baby (whose mother they did not know). The babies are six to twelve months old, and they are expressing joy, distress, and no particular emotion. The data were conclusive: strong responses in mirror neuron areas, in the insula and in limbic areas. Mothers are highly empathic subjects, and we were pleased to see such robust responses in this circuit, which, as depicted in figure 2, connects mirror neurons with the emotional brain centers and, in this instance, allows the empathic understanding of the babies' emotional states by virtue of simulating the observed facial expressions. Any other result would have sent us back to the drawing board in a state of shock.

What about the mother's comparative response to her own baby and another baby? Again, as we expected from the behavioral data, the neural circuit was more active while the mothers were watching the expressions of their own babies. But we also found something unexpected. Another region "outside" the previously known circuit was strongly activated while our moms were watching their own children, compared with somebody else's. It is called pre-SMA, and we know it is an important region for complex motor planning and motor sequencing—that is, for putting together a series of concatenated actions.

In the monkey, the homologue of pre-SMA is area F6. This is intriguing because F6 has strong anatomical connections with area F5, one of the areas of the monkey brain containing mirror neurons, as we know. Furthermore, data suggest that area F6 may control and modulate the activity of cells in area F5.[17] Therefore the strong response in pre-SMA in the mothers' brains suggests that when watching her own baby, a mother not only mirrors the emotions of the facial expressions of the baby but also activates a series of motor plans to interact with the baby in an effective way. After all, if a baby cries, it is hardly helpful for the mother to cry too! An effective interaction requires that the mother respond appropriately to console the baby. The high activity in pre-SMA while mothers are watching their own babies probably represents the simulated initiation of a series of appropriate actions in response to the emotional situation of the child.

Considering that the mothers in this experiment were, like all fMRI subjects, lying inertly inside a big, noisy machine, and that they were simply watching pictures of babies, these strong responses in pre-SMA are amazing. They suggest that the initial automatic mirroring of the facial expressions of the babies triggers a whole cascade of other automatic simulative brain responses that reenact interactions between mother and baby in real life. This constant automatic simulation and reenactment has the purpose of making us ready when action is really needed. This is probably especially true in the domain of empathy, where one of the defining elements is the ability to respond compassionately to another person's distress. In the case of maternal empathy, this ability surely reaches its highest possible expression.

FIVE

Facing Yourself

IS IT YOU OR IS IT ME?

Almost instinctively we humans tend to synchronize our movements. I fold my arms, you fold yours, I look at you, you look away, you look back, I look away, I look at you, you start a new sentence, you look at me, I start a new sentence—it's quite a minuet we're dancing! In videotaped situations such as I've described in earlier chapters, it is fascinating to observe. It also turns out that the more people like each other, the more they seem to imitate each other, and this makes sense too. This imitation and synchrony is the glue that binds us together. Thus my confident assertion that mirror neurons are integral to the requirement that we humans fit ourselves as smoothly as possible into our social context. We're all in this boat together, and mirror neurons help us make the best of it. We need them. They help us recognize the actions of other people, imitate other people, understand their intentions and their feelings. However, if we stop to think about it, these functions performed by mirror neurons create an interesting puzzle for neuroscientists investigating how the brain codes for "agency"—that is, the sense of being the *owner* of a given action. When I am grasping a cup of coffee while you are also grasping a cup of coffee, how does my brain distinguish my action from your action? This distinction is self-evident to us sitting here talking about it, but the great question for brain scientists is how, exactly, the brain makes it self-evident.

I believe the answer can be found in the very first experiment in which we investigated the role of mirror neurons in imitation, where we measured brain activity while subjects were performing and imitating hand movements. My discussion in chapter 2 did not mention the most surprising element in the results: the parietal operculum, an area of the brain that receives sensory information from the hands (did they open or close? was the touched object soft or sharp?), showed higher activity during imitation compared with the mere execution of the same movement.[1] This sounds like mirror neuron activity, doesn't it? During imitation, these neurons basically "add up" the activations from observation and from execution. The problem is that the parietal operculum is *not*, to our knowledge, an area with mirror neurons and has no known mirroring role. Indeed, the parietal operculum was not activated during action observation. The hand movements in the experiment were substantially identical, so the information these cells were receiving from the moving hands was also substantially identical. Why, then, the mirror neuron–like difference in activation? We were quite surprised.

Our investigation focused on the fact that the enhanced activity in the parietal operculum was localized in the right cerebral hemisphere, a very important region for the mental representation of our body and its extremities—our body schema, as it's called. (We know this because patients with lesions in this region may have severe disorders of body awareness. They may deny there's anything wrong with their paralyzed left arm, or claim that it's not actually theirs, but in fact belongs to a relative. They may even believe they have more than two arms.) The enhanced activation during imitation that occurs in this important region for establishing body awareness might be the brain's way of overcoming any possible confusion caused by mirror neurons, any tendency they might have to make us lose our sense of being the agents of our own actions. This is a way the brain has devised to reaffirm a sense of ownership of our own actions.

To this point in the book, I have argued that the major role of mirror neurons is to allow the understanding of the intentions and emotions of others, and thus to facilitate social behavior. They seem just about as "interested in" other people as they are in the self in whose brain they reside. Their firing pattern may give the impression that mirror neurons are not highly relevant to the construction of a sense of self. That might be your impression right now, anyway, and it would be a reasonable impression, but I am going to spend the rest of this chapter modifying it. Let me start with some theoretical considerations, of which the most intriguing might be the position of several authors (especially in the phenomenological tradition, as we will see in the last chapter) who have argued that we cannot and should not artificially separate self and other. They are "co-constituted," in the parlance. As the philosopher phenomenologist Dan Zahavi puts it, "They reciprocally illuminate one another, and can only be understood in their interconnection."[2] Perhaps odd at first blush, the argument quickly starts making more sense. How can we even think of "self" except in terms of the "other" that the self is not? Without self, it makes little sense to define an other, and without that other, it does not make a lot of sense to define the self. And how could mirror neurons not play a role here? These are the very brain cells that seem to index (with their neuronal firing pattern) this unavoidable relationship between self and other, the inevitable interdependence. Note carefully, however, that the firing rate of mirror neurons is not the same for actions of the self and actions of others. As we have seen time and again—in every experiment ever conducted on mirror neurons, in fact—there is a much stronger discharge for actions of the self than for actions of others. Thus mirror neurons embody both the interdependence of self and other—by firing for the actions of both—and the independence we simultaneously feel and require, by firing more powerfully for actions of the self.

My theory of how mirror neurons become the neural glue between self and other starts with the development of mirror neurons in the infant brain. Although there are no empirical data available yet, a very likely scenario is not hard to surmise. Baby smiles, the parent smiles in response. Two minutes later, baby smiles again, the parent smiles again. Thanks to the imitative behavior of the parent, the baby's brain can associate the motor plan necessary to smile and the sight of the smiling face. Therefore—presto! Mirror neurons for a smiling face are born. The next time the baby sees somebody else's smile, the neural activity associated with the motor plan for smiling is evoked in the baby's brain, *simulating* a smile. If this account of how mirror neurons are initially shaped in our brain is true—and I believe it almost certainly is—then "self" and "other" are inextricably blended in mirror neurons. Indeed, according to this account, mirror neurons in the infant brain *are formed by the interactions between self and other*. This is the key concept to keep in mind for understanding the role of mirror neurons in human social behavior. It makes sense that later in life, we use these very same brain cells to understand the mental states of other people. But it also makes sense that we use the same cells to build a sense of self, since these cells originate early in life when other people's behavior is the reflection of our *own* behavior. In other people, we see ourselves with mirror neurons.

Indeed, another point in favor of the link between self and other, imitation and mirror neurons, comes from empirical data. A developmental study investigated spontaneous imitation in pairs of children. With some of these pairs, both kids had acquired the ability to recognize themselves in front of a mirror; with some of the pairs, neither kid had yet acquired this ability. The results were clear. The pairs of kids with the ability to recognize themselves in front of a mirror imitated each other much more than the pairs of kids who did not yet have the mirror recognition ability.[3]

Self-recognition and imitation go together because our mirror neurons are born when the "other" imitates the "self" early in life. Mirror neurons are the neural consequence of this early motor synchrony between self and other, and they become the neural elements that code the actors of this synchrony (the self and the other, obviously). Of course, we must have some mirror neurons at birth, given Meltzoff's data on imitation in infants. However, my argument is based on the assumption that the mirror neuron system is largely shaped by imitative interactions between self and other, especially early in life (although I believe that the experience of being imitated can shape mirror neurons also later in life, as we will see in the next chapter). According to my theory, it makes sense that the pair of kids who could self-recognize were also the ones who imitated more. The same neurons—mirror neurons—are concerned with both, and when they can implement one function (self-recognition), they can implement the other function (imitation) too. But what do scientists mean with the term "self-recognition"?

THE MIRROR RECOGNITION TEST

The concept of the self is a highly complex one. Several factors participate in its creation, which is typically a problem for experimenters who need to reduce the complexity of a phenomenon to manageable dimensions. On this question, fortunately, we now have a relatively simple experimental paradigm that allows us to search for some aspect of self-awareness—an important aspect, we believe—even in very young children, even in animals. We call this the mirror recognition test. (More mirrors! If you're starting to feel that there are too many in this book, I can't really blame you, nor can I do much about it. Certainly not here. This is an important, fascinating test.) It was devised by Gordon Gallup, professor of psychology at the State University of New York at Albany, in the late 1960s. As was the case with the discovery of mirror neurons, some serendipity came into play. As a graduate student, Gallup was taking a course that required a research project. He takes up the story: "I found myself contemplating what to do one morning as I was shaving in front of the mirror. While confronting my image in the mirror I began to wonder if other species might be capable of recognizing their own reflections, and how you could test for this."[4]

As Gallup soon learned, scientists had been using mirrors in just this way since the mid-nineteenth century. Charles Darwin was among the first, testing his own children (they eventually passed) and correctly proposing that such ability is a sign of higher intellect. Testing this hypothesis, Darwin placed the mirror in front of two orangutans in the Zoological Gardens in London. He thought the two animals behaved as if they were facing two other animals rather than their own reflected images. For about a century, other scientists used the mirror in similar investigations of both animals and infants, but their assessment of the subjects' behavior in front of the mirror was, like Darwin's, only a descriptive one, with all the shortcomings that a purely descriptive method inherently has. Simply put, those shortcomings mainly include no objective standard of evaluation, as well as unrecognized bias on the part of the investigator. Gordon Gallup took this old, incompletely realized research idea and formalized it with beautiful simplicity.

His first test subjects were chimpanzees, and he initially did nothing more than to simply observe their spontaneous behavior in front of a mirror. In this regard alone Gallup's work was an important step forward. For the actual test, the chimps had to be familiar with the mirror. Otherwise, the element of surprise would skew all results. They also had to have the opportunity to learn about the mirrors, and he needed to acquire a good sense of their spontaneous behavior in the presence of the mirror as it developed over time. This initial phase of the study took several days. In the second phase, the animals were anesthetized and an odorless dye was used to mark their foreheads. The point is straightforward: the mark could not be viewed by the animals directly, only through the mirror. When the animals woke up from the anesthesia, Gallup did not rush the mirror into place. He had to study their behavior to determine whether they were able to feel it or smell the mark. The chimpanzees showed no such awareness, made no attempt to touch the mark, and did nothing out of the ordinary at all. Only now did Gallup introduce the mirror again—and a sudden change in the behavior of the animals was immediately noticeable. The chimpanzees were frequently touching the mark, examining it closely, and repeatedly using the mirror! Proof positive: they knew they were looking at themselves in the mirror. With the simple idea of marking the forehead of the animal, Gallup had devised an effective and objective way of testing self-recognition in animals. In principle, the whole test could be reduced to two numbers: the number of times the animal touches its forehead, with and without the mark.

Gallup also ran a necessary control experiment. Some chimpanzees who had never been exposed to a mirror were anesthetized and marked on their foreheads with the odorless dye. Given that these animals had never seen their own faces, Gallup predicted that they were not going to show the dramatic mark-directed behavior displayed by the previous group of chimpanzees. Indeed, this second group of chimpanzees basically ignored the mark even when in front of the mirror.[5] Why? The mark was not remarkable. They had no way to know it had not always been there.

The mirror recognition test quickly became a widely used and highly popular tool in animal cognition research. Monkeys fail the test. They see their reflection as another monkey, try to play with him, and when this fails, look *behind* the mirror to try to find out what in the world is going on. Too bad we can't perform single-cell experiments under these circumstances—their mirror neurons are, presumably, going crazy! Darwin, it turns out, was wrong. Orangutans, especially those raised in a human environment, pass the test.[6] Puzzlingly, most gorillas do not pass the mirror recognition test. The few who do were raised in a rich human environment.

The fact that the social context is critical in developing

self-recognition abilities in apes is telling. Isolation seems to inhibit the ability to develop self-recognition; rich social context facilitates it. What is the main difference between the two environments? The presence of *others*—the continuous relations and interactions one must have with other individuals. Mirror neurons fire when we observe actions and when we perform those same actions. In short, when we (and apes) look at others, we find both them *and ourselves*. The link between social environment and a sense of self is strongly inferred.

If this is so, we might expect that other highly communicative animals with social skills would show signs of self-recognition in the mirror recognition test. Indeed, dolphins seem to have self-recognition, although it is difficult to study mark-directed behavior in animals that do not have limbs. In a recent study, dolphins marked on various parts of their bodies spent more time in front of a large mirror placed under the water, compared with when their bodies were not marked. Also, the dolphins twisted and angled their bodies in a way that suggested they were trying to look at their marks. Considering that imitative and empathic behavior has been associated with dolphins, the evidence for some form of self-recognition in these animals shows yet a further link between imitation, empathy, and sense of self.[7]

Elephants have also been associated with social complexity and empathic behavior. Can elephants also self-recognize? An early study suggested that they cannot. However, testing elephants with mirrors is not a trivial enterprise from a practical standpoint. A very large mirror is obviously required! In a recent study using a mirror 2.5 meters tall and elephant-resistant as well, scientists did obtain evidence for mirror recognition.

The capacity to self-recognize in primates, dolphins, and elephants—lineages that separated a long time ago in evolutionary terms—demonstrates convergent evolution that is probably due to the interaction between biological factors and environmental ones.[8] Most likely, the highly sophisticated social interactions displayed by these lineages are expressions of both biological predisposition and the role of experience in shaping behavior. Indeed, all of these animals have rich mother-infant interactions that last quite a long time. As I discussed earlier in this chapter, parent-infant reciprocal imitation is very likely a key form of experiential shaping of mirror neurons. Rich and long mother-infant interactions may be one of the factors favoring convergent evolutions between primates, dolphins, and elephants, thereby facilitating the shaping of mirror neurons and of a sense of self.

The mirror recognition test has naturally also been widely used with children—without the anesthesia. Instead, the investigators have to use some tricks, either waiting until the child falls asleep or distracting the child while applying the mark. The results of these studies are fascinating. Nearing one year in age, children can spend an enormous amount of time in front of a mirror but consistently fail to pass the test. For the children, as with monkeys, it is as if the image reflected in the mirror is not so much their own, but rather the image of another child they are playing with, as Julian Keenan, Gordon Gallup, and Dean Falk put it in their book *The Face in the Mirror*.

Toward the end of the second year, children catch on and consistently show behavior directed at the mark. And along with this behavior, other behaviors that clearly show a social awareness also appear. For instance, the children show the first signs of embarrassment.[9] Embarrassment requires at least some rudimentary sense of social norms, which are derived from daily interactions with other people. When children are embarrassed, they are embarrassed in front of other people.

As demonstrated above, it is quite likely that mirror neurons play a major role during social interactions very early in life, starting with the first interactions with Mom and Dad. If I am correct that social interactions shape the ability to develop a sense of self, as both the animal and the developmental data reviewed in this section of the chapter suggest, then it is also likely that mirror neurons are involved in self-recognition. Let's consider the brain data supporting this hypothesis.

ANOTHER ME

Some years ago, Lucina Uddin, a graduate student in psychology at UCLA, told me that she wanted to use both fMRI and TMS to better understand the neural correlates of self-recognition. She had already performed a couple of experiments in Eran Zaidel's lab. Zaidel was one of the first scientists to look at the neural correlates of the self. With his wife, Dahlia, and their mentor, Roger Sperry, he investigated how the two cerebral hemispheres recognize faces, and he began by examining patients in which the two hemispheres had been *disconnected* in an attempt to reduce intractable epilepsy.

The neurosurgeon was Joe Bogen, who had first performed this procedure—specifically, "sectioning" the corpus callosum—on epileptics in the early 1960s. The corpus callosum is a very large bundle of brain fibers that connects the left and the right cerebral hemispheres. By cutting it, Bogen effectively stopped the "spread" of epileptic activity from an initial area on one side of the brain to the other side. The procedure did help the patients. Bogen also sectioned the anterior and posterior commissures—two other, much smaller bundles of brain fibers that also connect the left and right hemispheres. For all practical purposes, these patients now had two separate brains, and they became known as split-brain patients. The series of operations became known as the West Coast series, because Bogen performed them in Southern California. At that time, Roger Sperry, who was at Caltech in Pasadena, started a systematic investigation of the psychological functions of each separated cerebral hemisphere in Bogen's novel split-brain patients. For this investigation Sperry was eventually awarded the Nobel Prize for physiology and medicine.

The experiment that Sperry and the two Zaidels performed with these split-brain patients took advantage of the fact that, owing to the anatomical organization of the visual system in the brain, any visual stimulus that appears in the *left* side of the field of vision goes to the *right* cerebral hemisphere, and vice versa. They presented to these patients a series of visual stimuli in one field of vision or the other, and therefore to one hemisphere or the other. Among these stimuli was the face of the patient. General belief at the time held that the left cerebral hemisphere would be the only one able to recognize the patient's own face, since it is the hemisphere with more developed language faculty. This belief was based on the assumption that in order to recognize one's own face, one needs to verbalize (at least covertly) this whole process. However, Sperry and the two Zaidels discovered that both hemispheres could recognize the patient's own face, debunking the generally held belief that only the left hemisphere could enable self-recognition.[10]

Working in Zaidel's lab at UCLA, Lucina Uddin used a slightly different approach to the same issue, an approach developed by the cognitive neuroscientist Julian Keenan at Harvard University. The visual stimuli she flashed to either the left or right field of vision were also faces—of a sort. They were actually a series of morphs of the patient's face and another face, morphing in incremental steps of 10 percent from all-self to all-other face. The task for the patient was to report whether each morph was more her own face, or more the other face. Even though this approach was quite different from Sperry's initial study, Lucina's results were in fact very similar, suggesting that both the left and the right hemispheres can recognize one's own face.[11]

Now she wanted to use a variety of experimental methods in my lab to further explore the relations between the self and the brain, all serving the same fundamental question: Which regions of the brain are critical for self-recognition? The most logical approach in her position was to use an experimental paradigm similar to the one with the series of morphing faces, this time placing healthy volunteers in an fMRI scanner as Lucina measured how different brain areas were activated by the sequence of morphing faces. She contacted me mainly because I am on the faculty of the Brain Mapping Center at UCLA and have collaborated with her mentor, Eran Zaidel. In any event, she wasn't thinking about mirror neurons—but I definitely was.

My hypothesis that mirror neurons are formed by the interactions between self and other (baby smiles, caregiver smiles too) and the developmental data on self-recognition and imitation predicted the involvement of mirror neurons in self-recognition. I was not an expert in self-awareness, however, and I needed one in order to move ahead with experiments. Lucina was the ideal trainee: she knew practically everything about these "self and the brain" issues, from evolutionary to developmental perspectives, from philosophical discussions to the scant data available at that time. She needed training only in brain imaging. This I could provide, and I jumped at the occasion and eventually became co-mentor for her doctoral studies. I hope that at the end of her doctoral years Lucina has learned something from me with regard to brain imaging, because I can definitely say that I have learned a lot from her with regard to the self.

One of the problems to solve when designing an experiment about self-recognition is the fact that we typically look at our own face several times every day. Our face is highly familiar to us, even if it changes as the years roll by. Therefore any non-self stimuli used in the experiment would have to be highly familiar to the subject too; otherwise we would risk finding experimental results having more to do with changes in visual familiarity than with self-recognition. Previous investigators in this area had recognized this problem and solved it by using, for the non-self faces, those of famous people: Marilyn Monroe, Albert Einstein, Bill Clinton. Almost every adult in the Western world knows these faces. However, your typical research subject in the brain imaging lab—a college student or graduate student—does not spend time with these people, so they have small social meaning for the subjects. In contrast, the subjects' own faces would have high social relevance. For all of us, not just the subjects in these tests, our own face is not simply the one reflected in the mirror. It is also the face people see when we interact with other people, the face that communicates our own emotions to them. One of the reasons people obsess about their appearance, and about their face in particular, is its high social valence, as we say. For this reason, Lucina introduced a novel element in her brain imaging experiment on self-face recognition: the non-self face, rather than being a highly familiar but socially irrelevant celebrity face, was instead a highly familiar *and* highly socially relevant face—the subject's best friend. I believe that this idea was key to the ultimate success of Lucina's experimental design.

I remember when I saw her results for the first time—the comparison of brain activity while subjects were looking at the sequence of pictures morphing incrementally from their own to their best friend's face. I was stunned. The two areas that stood out clearly for self-recognition compared with recognition of the best friend—one in the frontal lobe of the right hemisphere, one in the parietal lobe—are also the mirror neuron areas of that hemisphere. Lucina had beautifully imaged the whole mirror neuron system of the right hemisphere (figure 1, p. 62). A few weeks after I saw these wonderful results, I was at Dartmouth College as part of the faculty of a summer course in cognitive neuroscience, widely known as Brain Camp. Giacomo Rizzolatti was also there; in fact, he was one of the organizers that year. One day while we were sitting in the lobby of the hotel, working on the paper on mirror neurons and understanding the intentions of others (discussed at the end of chapter 2), I showed Giacomo the images from Lucina's experiment.

"What do you think it is?" I asked.

"It is definitely the mirror neuron system in the right hemisphere. But I have rarely seen such a strong mirror neuron activity in the right hemisphere only," Giacomo added, and then asked, "What kind of task are the subjects doing?"

When I explained the self-recognition morphing experiment, he nodded. The results made sense to him. I realized that I was not chasing a fantasy. If it made sense to Giacomo and it made sense to me, and as we had never actually talked about self-recognition and mirror neurons before that day, it was a good sign. But why should mirror neuron areas in the human brain be incrementally activated as we incrementally recognize our own face in a series of morphed photographs? Aren't mirror neurons the cells that fire when we perform an action, or see someone else perform that action, or imitate that action? Why should they fire at all when we see an *unmoving* face? The answer, I believe, is this: first of all, it is well known that the human brain responds to static stimuli that imply motion *as if* those stimuli were moving. For instance, the main human brain area that responds to motion (we brain scientists call it area MT) responds also to still pictures of animals in motion—running or jumping—and even to still pictures of natural scenes implying motion, such as ocean waves. Likewise, the human mirror neuron system responds to still pictures implying actions—for instance, a hand in the middle of a grasping action.[12] Now, let's go back to Lucina's experiment. She used still pictures of faces. It turns out that the perception of a face almost invariably implies motion. It is very difficult to look at a face and not to think about it in motion, making facial expressions. The activation of mirror neurons when watching a face is therefore not so surprising after all.

We now have to explain the increased activation in the mirror neuron system that Lucina observed for self, compared with other. We have seen that mirror neurons—through a simulation mechanism—map the actions of the *other* onto the *self*. They make the other "another self," as Gallese says. When we see our own picture, there are actually two selves facing each other. The "perceived self" is the one in the picture, while the "perceiving self" is the one *watching the picture*. Mirror neurons in the brain of the perceiving self process the perceived self as the *other*, and again implement the mapping of the other—in this case those implied facial expressions—onto the self. But now those implied facial expressions of the "perceived self" already belong to the motor repertoire of the "perceiving self." Thus the simulative process implemented by mirror neurons is highly facilitated and results in higher activity in the human mirror neuron system.

This explanation made a lot of sense to me, and it is, obviously, the explanation we put in the paper that reported Lucina's findings.[13] However, I thought it was important to test our hypothesis more directly. After all, we cannot forget that the data we had obtained from the fMRI scanner were only correlative data. They did not tell us whether there was a *causal* link between the activation of those brain areas and the ability to self-recognize. A powerful tool to investigate causal links between a brain area and a given function is transcranial magnetic stimulation, as we have seen. The transient disruption of brain activity in a localized area induced by the TMS apparatus reveals whether this area is necessary for the task performed by the subjects. After Lucina had finished her fMRI experiment, I urged her to use the TMS machine to determine whether those activated areas are really necessary for self-recognition. Thankfully, Lucina is a good listener (not all students are).

ZAPPING THE SELF

Lucina was able to track down her "morphing" subjects, and most agreed to be the subjects in a new experiment using the copper coil. As we have just seen, her first experiment had revealed two areas in the right cerebral hemisphere that are also in "mirror neuron territory" as most active for self-recognition. Which one should she focus on in this new experiment? To help answer this question, she reviewed the very rare cases in the literature of neurological patients who show a deficit in recognizing their own face while they are in front of a mirror. We call this deficit the mirror sign. If these patients don't see themselves in the mirror, who or what do they see? Someone else (in a way)! One patient described another person who looked exactly like her. A second patient described seeing a young girl who looked like her. A third patient saw his double, his look-alike. And a fourth patient simply saw somebody who followed him around. Otherwise, these patients were perfectly able to identify other people in the mirror, and they were able to use the mirror correctly, for grooming purposes, for instance. Their deficit was strictly self-recognition.

In none of the four patients did the brain lesions unequivocally pinpoint a specific brain region as responsible for the self-recognition deficit. However, all did show more involvement of the right hemisphere than of the left hemisphere; in the right hemisphere, they showed more involvement of brain areas toward the back, in the parietal lobe.[14] Among these areas, the supramarginal gyrus is anatomically very close to the areas damaged in the mirror-sign patients. It seemed like an obvious candidate as a critical brain area for self-face recognition.

No two brains have the same size and shape and internal structure, nor does the activation of the same brain area in different individuals ever look exactly the same. Lucina used the functional brain imaging data of each individual and the frameless stereotaxy system with the infrared camera described in chapter 3 to place the coil and stimulate her target with precision. That target was about the size of a square centimeter of brain surface, just underneath the copper coil placed over the skull. Lucina chose to use the low-frequency repetitive stimulation, which, in simplest terms, zaps the target with one magnetic pulse per second for a prolonged period of time, typically twenty minutes, transiently reducing the activity of the targeted neurons for approximately half an hour after the end of the stimulation[15]—in this experiment, the neurons that were quite active in self-recognition in the morphing-photographs experiment. Lucina's prediction was straightforward: if the activation of this area is essential to the ability to self-recognize, then her subjects should have reduced self-recognition ability right after the TMS stimulation. This is exactly what she observed: significantly reduced performance. The percentage of "self" in the morphed photographs had to be higher in order for the subjects to make the self-identification.

Obviously, Lucina also needed to stimulate another brain area as a control for nonspecific effects owing to magnetic stimulation. The observed reduction in performance could have been due to a variety of factors that had little to do with the zapping of the stimulated brain area. For this control experiment, she used the corresponding brain area in the left hemisphere. Since her original imaging data had shown no activity for self-face recognition in this left supramarginal gyrus, it seemed like a good choice. Stimulation of this area should not affect self-recognition capabilities. If it did, suspicion would be directed at some other factor. But it did not. The performance of Lucina's subjects before and after stimulation of the *left* supramarginal gyrus was substantially identical, thereby confirming that the deficit in self-face recognition induced by stimulating the *right* supramarginal gyrus really was due to that specific disruption. Although the involvement of this brain area in self-recognition had been shown by Lucina's previous study and by other studies, in particular the work of Julian Keenan, this TMS experiment demonstrated for the first time a fairly precise causal relation between a specific human brain area and the capacity to recognize oneself.[16] In addition, other kinds of neurological data had suggested this association. It is well known that patients with right hemisphere lesions may develop a neurological phenomenon called asomatognosia. These patients cannot recognize a body part as their own, typically a paralyzed left arm or hand. They believe, for example, that it belongs to one of their relatives. Comparing the location of lesions in patients with right hemisphere damage, with or without asomatognosia, the neurologist Todd Feinberg found that all the patients with this deficit had lesions in the supramarginal gyrus, whereas none of the patients without this deficit had such lesions.[17]

Obviously, neither Lucina's brain imaging and magnetic stimulation data nor any other noninvasive work can demonstrate with certainty that mirror neurons, specifically, are activated during self-face recognition and that the disruption of *their* activity is responsible for the self-recognition deficit. This technology does not have the resolution of looking at the activity of single cells. However, as we have seen in chapter 2, the area stimulated by Lucina is an area with clear mirror neuron properties. I have argued earlier in this chapter that the interactions between "self" and "other" (baby smiles, mother smiles too) shape mirror neurons early in life. Mirror neurons link profoundly self and other. I would even argue that self and other are blended in mirror neurons as they are blended in Lucina's morphed faces. The well-confirmed "interest" of mirror neurons with the other must in some way entail interest with the self. Lucina's data comprise the empirical evidence that most strongly supports this concept. Her findings show the biological roots of intersubjectivity. Unfortunately, philosophical and ideological individualistic positions especially dominant in our Western culture have made us blind to the fundamentally intersubjective nature of our own brains. I believe that the neuroscientific work on mirror neurons proves it. I will explore the theoretical implications of this critical aspect of our brains in the last chapter.

But how abstract is the relationship between mirror neurons and the self? Another member of my lab, Jonas Kaplan, performed a brain imaging experiment to test whether the human mirror neuron areas are activated during a self-recognition task involving the *voice* of the subjects. Jonas has a strong cognitive and philosophical background, including a special interest in Eastern philosophy. He also has many talents, including playing the sitar, mostly Indian music but also in a rock band. Impressed (as everyone in the lab was) with Lucina's experiment, Jonas wanted to see whether the areas Lucina had identified in her self-recognition task with a visual stimulus—the face—would also discriminate between a person's own voice and that of somebody else. If this turned out to be the case, Jonas thought, it would suggest that these brain areas are concerned with relatively abstract representations of the self. For his study, Jonas followed Lucina's lead and chose for his comparison test the voice of the best friend of each subject. But there was a huge catch: there is no way to morph voices. Given this fact, Jonas decided to use unmorphed pictures and unmorphed voices. The subjects in the experiment watched their own faces as they heard their own voices, and then watched their best friends' faces and heard their voices. A solid plan, but Jonas had to solve another problem, the well-known fact that the sound of our voice while speaking is fundamentally different from the sound of our taped voice. The physiological reason is that when we speak, our voice is transmitted to the ear not only through air but also through body tissues, mostly bone. So Jonas filtered the taped voices in such a way that the taped and live voices for each subject were very similar—clever work.

Why was Jonas so interested in checking whether the human mirror neuron areas are activated in self-recognition tasks involving both face and *voice*? Recall that mirror neurons in monkeys respond to sounds—in the experiments described in chapter 1, these were "action sounds" associated with observed actions, such as tearing open a peanut, ripping a piece of paper. In chapter 3 I have discussed the analogous results in brain imaging experiments with humans, as well as other experiments demonstrating that humans show mirroring phenomena with regard to *speech* sounds. Clearly, then, these mirror neuron areas are multimodal, responding to both visual and auditory stimulation, and we would therefore expect that the specific mirror neuron areas activated during Lucina's self-recognition experiment with photographs should also activate during Jonas's closely parallel experiment with voices. Their "failure" to do so in Jonas's experiment would be difficult to reconcile with the hypothesis of a role of mirror neurons in self-recognition.

I am pleased to report that there was no such failure. The same areas that activated for the pictures of the subjects also activated for their voices, demonstrating that mirror neurons code for multiple "self-related" stimuli, confirming their importance for self-recognition (and a rather abstract representation of the self at that).

TWO SIDES OF THE SAME COIN

We have seen from animal studies that the development of a sense of self is facilitated by rich social contexts. From developmental studies, we have seen that self-awareness corresponds developmentally with forms of social behavior, from imitating others to expressing socially oriented emotions such as embarrassment. Finally, we have just seen that areas of the human brain known to have mirror neuron properties are activated in self-recognition tasks involving both face and voice, and that the transient disruption of their activity by TMS induces self-recognition deficits. All these data, together with the theoretical considerations discussed at the beginning of the chapter, suggest that mirror neurons are important for my two-sides-of-a-coin analogy, in which one side is the self, the other side, well, the other.

The attempt to separate the two sides of a coin makes little sense. One ends up with not a coin, but a worthless piece of metal. Unfortunately, Western culture is dominated by an individualistic, solipsistic framework that has taken for granted the assumption of a complete separation between self and other. We are entrenched in this idea that any suggestion of an interdependence of self and other may sound not just counterintuitive to us, but difficult, if not impossible, to accept. Against this dominant view, mirror neurons put the self and the other back together again. Their neural activity reminds us of the *primary* intersubjectivity,[18] which is, of course, the early interactive capacities of babies displayed and developed in mother-baby and father-baby interactions. Are mirror neurons formed during and shaped by this primary intersubjectivity? I believe so. Although it is likely that some mirror neurons are functioning very early in life and facilitate the earliest interactions, I believe that most of our mirror neuron system is actually formed during the months and years of such interactions. The shaping of mirror neurons in the baby's brain is especially likely to happen during reciprocal imitation, as we saw in the consideration of smiling. If mirror neurons are actually shaped in our brain by the coordinated activities of mother and father and baby, then these cells not only embody both self and other, but start doing so at a time when the baby has more of an undifferentiated sense of *us* (mother-baby or father-baby) than any sense of an independent *self*, before the baby can pass the mirror recognition test. From this primary "us," however, the baby slowly but surely comes to perceive the other naturally and directly, and obviously without any complex inference; it proceeds to carve out a proper sense of self and other. How? With the help of a special type of mirror neurons, which I called super mirror neurons. We'll look into these cells in chapter 7. Throughout life, then, the activity of mirror neurons continues to be the neural signature of this sense of us to which both self and other belong.

SIX

Broken Mirrors

BABY MIRRORS

As I noted in the first pages of this book, a main reason Giacomo Rizzolatti and his colleagues in Parma began their experiments on the neurophysiological mechanisms for motor control in the macaque monkeys was the hope that their research would eventually lead to breakthroughs that could help humans recover motor functions after brain damage. They were neither looking for nor expecting to find mirror neurons. But find them they did, and their discovery has opened up a whole new realm of hope. As the evidence mounted for the role of mirror neurons in social learning and social behavior, one of the great dreams we all began to indulge in was that our research would teach us more about social disorders, such as autism, with effective treatments to follow.

Understanding how the mirror neuron system develops early in life is obviously very important in relation to autism, a disorder affecting roughly one out of a thousand children. Autism is diagnosed within the second year of life, as the child begins to show severe deficits in social relations. Several labs are exploring the hypothesis that a dysfunction of the mirror neuron system is responsible for autism, and some scientists are already exploring the implications for treatment. One of the obvious strategies suggested by the mirror neuron hypothesis is the use of imitation in treatment. Indeed, there are already scientific reports showing some beneficial effects of imitation-based treatments of children with autism. This is very exciting—and I guess it is why I am getting ahead of myself. First I should explain what we know—or at least speculate, since there are not a lot of empirical data yet—about mirror neurons in typical development, then discuss the data suggesting a dysfunction of mirror neurons in autism, and finally look at the promising new treatments.

We saw in chapter 2 that infants can imitate some rudimentary facial and hand movements. This capacity for imitation is probably supported by mirror neurons. Obviously, we do not have (and presumably will never have) direct observation of single-cell activity in human infants to prove the point, but some brain imaging data have recently emerged that support the existence of a mirror neuron system in the infant brain. The two main technologies, fMRI and TMS, employed in the lab at UCLA were not designed with infants in mind, and it is difficult to obtain good experimental data when these techniques are applied to babies. It has been tried, but not very successfully. However, certain "optical imaging" techniques are well suited for the infant brain because they do not require the subject to lie still while inside a large machine.

The main idea here is simple: When we direct light toward any object, some of the light is absorbed, some is reflected. In the case of the brain, the physiological processes in the active brain actually change the amount of the light that is absorbed or reflected. By measuring these changes, optical brain imagers can measure brain activity while the subject—even an infant—is performing the task called for by the experiment. One of these techniques is near infrared spectroscopy (NIRS), which uses light that is near infrared to study brain activity in babies in highly naturalistic settings. In one recent experiment, two Japanese brain imagers used NIRS to study brain activity in infants just six and seven months old.[1] The probes emitting and detecting the light (called optodes) were fitted on the head of the willing but unwitting subject using a soft headband and a special holder made specifically for the experiment. A parent was on hand holding the infant on his or her lap. With everything ready to go, the experimenters videotaped the infants' movements while they played with toys. Subsequently, they compared activity in the brain during periods in which the babies moved a lot against activity during periods of relatively little movement. This comparison told them which areas of the infants' brains were motor areas. Using this information, they then placed several optodes directly over these motor areas to determine whether or not they would be activated when the infants watched somebody else making an action. If these motor areas were also active while the babies were simply watching the actions of somebody else, then that brain activity was most likely due to mirror neurons.

In this observation-only phase of the experiment, the babies watched three different movements: a woman playing with a toy, a toy moving "by itself" (with the experimenter using a long cord for manipulation), and a ball swinging like a pendulum (following physical laws, since it was hanging from the ceiling). Some babies watched these experimental conditions live; others watched the same conditions through a television monitor. The spectroscopy equipment was rolling, and the experimenters were also tracking the movements of the babies as they watched the scenes. In their subsequent analysis, the brain imagers did not analyze periods of excessive motion of the babies, and when they compared the experimental conditions against each other, they took care in comparing only those periods in which each subject was exhibiting roughly equivalent movement of his or her own. The scientists also controlled for attention, selecting those sequences when a given baby was paying attention and watching the stimuli throughout the experiment. Obviously, some of these six- or seven-month-old subjects did not manage to complete all phases of the experiment. However, about two-thirds did, and the results were very instructive. The motor areas of the babies' brains were activated when they watched the woman playing with the toy, but not when they watched the toy moving independently—clearly a suggestion (a strong suggestion) that mirror neurons are functioning well in young infants. Furthermore, the activation in these motor areas was higher when the babies were watching live actions compared with actions displayed by a monitor. This is a classic finding in mirror neuron research. Remember that in monkeys there is a strong discharge in mirror neurons when observing live action, but practically no response when viewing the same scene on a computer monitor. In humans, mirror neuron areas do respond to actions displayed by the monitor, but not as strongly as for live actions. The findings of the Japanese NIRS study were right in line.

Now we turn to babies one year old, but in a different experimental design. To understand this setup, we need to know that when we adults watch others moving objects—say, placing toys into a bucket—our eyes anticipate where the toys will be placed. We look at the bucket *before* the observed hand with the toy reaches it. This ability to "predict" with our eyes where objects will be placed by other people probably comes from our mirror system. Why? When we move those objects ourselves, our eyes do the same thing: we look at the bucket before our hand places the toys in the bucket, anticipating our own actions.[2] Infants six months old do *not* anticipate with their gaze where somebody else's hand is going to place the toy. In sharp contrast, year-old babies do this as if they were adults. Again, this ability probably originates from mirror neurons. If the toy is seemingly self-propelled, thanks to an experimental trick, the year-old babies cannot anticipate with their gaze where the toy will be placed. (Likewise adults! Watching a self-propelled toy heading for the bucket, we do not anticipate with our gaze where it will end up.[3] Our understanding of mirror neurons predicts this "discrepancy."

With the hand holding the toy, mirror neurons can code for intention; without the hand in the picture, they cannot.)

At six months of age, we cannot predict where the hand is taking the toy. At one year, we can. Clearly, mirror neurons *learn* to predict the actions of other people. This ability was not present at birth. This is another example of how the mirror neuron system may be shaped by experience.

THE TEEN BRAIN

If the mirror neuron system is very important in early childhood development, what a role it must play for older children! Given that it is such an important neural system for social behavior, how could it be less than critical as kids move toward their teenage years, when the whole of life sometimes seems to be pretty much defined through social networks and behavior. It is therefore critical to map the whole developmental trajectory, including in older children.

In Montreal, a group led by Hugo Théoret at the Montreal Neurological Institute is currently using electroencephalography to study the mirror neuron system in older kids. With EEG, electrodes placed over the scalp of the subjects record the electrical activity emanating from the surface of the brain. Using this technology to look at mirror neuron activity specifically, Théoret and his colleagues were monitoring something called mu rhythm. In simplest terms, the mu rhythm is the expression of oscillatory electrical activity that can be recorded over the central motor regions of the brain.

When we move our hands, say, the mu rhythm is *reduced*, or "suppressed," in the jargon of neuroscience. This inverse correlation between the mu rhythm and motor activity in the brain comes in very handy for neuroscientists. Mu rhythm suppression is a clear index of motor activity in the brain. Now, what should happen to the mu rhythm when we just observe the movements of others? Without knowledge of the mirror neuron system, we might be inclined to predict that the mu rhythm would not be suppressed. We're not moving, after all. But given our knowledge of mirror neurons, we are not surprised to learn that, indeed, simply watching other people perform actions also suppresses the mu rhythm in the brain.

This phenomenon had been discovered some years earlier by Riitta Hari and Giacomo Rizzolatti, using yet another brain imaging technique, magnetoencephalography (MEG). While TMS uses the copper coil to induce an artificial magnetic field to work its magic, MEG uses an impressive array of approximately three hundred sensors to pick up the much lower (infinitesimally low, really) magnetic fields that are spontaneously created on the surface of the brain by the electrical activity therein. The detected magnetic fields are mostly created by the activity of neurons on the bumps on the surface of the brain called gyri. Activity from the fissures of the brain (the sulci) is more difficult to measure with MEG. However, it remains a major brain imaging technique, primarily because its very high temporal resolution allows MEG specialists to discriminate neural responses on the order of few milliseconds. Let me give you an example. When we hear the telephone ringing and we walk toward it, the areas of our brain that respond to sounds activate before the areas that control walking. By using MEG to look at the temporal progression of activation in different brain areas, the specialist can figure out the communication going back and forth between brain areas.

Hari and Rizzolatti's discovery that the mu rhythm is suppressed both when we make an action and when we see somebody else doing so gave us yet another important biomarker of mirror neuron activity in the human brain.[4] Following their experiment, Hugo Théoret and Jean-François Lepage geared up to study mu rhythm suppression in normal, healthy children between four and eleven years old. For these experiments, the children either grasped an object or observed somebody else grasping the same object. Using EEG, Lepage and Théoret found mu suppression during both the execution and the mere observation of the grasping action, thus demonstrating what no one doubted: functioning mirror neurons in older children. (Shirley Fecteau, a student of Théoret's, also had the opportunity to study the EEG activity of the central motor areas of a three-year-old epileptic child, and mu suppression was recorded for both action and observation, in this case drawing a picture.) A key question these studies did not address, however, is how closely the mirror neuron system is related to the social competence and empathy of the children.

Investigating this question at UCLA, we have chosen to look at probably the most tumultuous period in human development—that is, adolescence. (I've mentioned that I am the parent of an eleven-year-old girl. I'm trying to steel myself for the next few years.) In the lab, we have recruited a large group of prepubescent children, and we plan to follow them (not literally, but with our imaging equipment) until they are fifteen. This longitudinal study is currently under way, but the first visit of these children has already been completed, so we have been able to look at how activity in the mirror neuron system of typically developing ten-year-old children is related to their social competence. This particular study was led by Mirella Dapretto, a developmental psychologist expert in pediatric brain imaging and autism who also happens to be my wife.

For this first study, Mirella used fMRI for a social mirroring task that required the children to observe and imitate facial emotional expressions. Recall from chapter 4 that we used this same task involving facial expressions to investigate the functional link between the mirror neuron system and the emotional brain centers in the limbic system (the link provided by the insula, as it turns out). Rather than simply look at the mirror neuron system, as the studies with the younger children had already done, Mirella wanted to look at these same functional connections between the mirror neuron system and the emotional brain centers, connections that presumably allow social mirroring and the understanding of the emotional states of other people, thus facilitating empathy.

To test her hypothesis, Mirella acquired brain imaging data on the children and also assessed their empathic abilities and their interpersonal competence. She measured the empathic abilities with the Interpersonal Reactivity Index, a well-tested scale composed of four subscales, two measuring cognitive empathy, the other two emotional empathy. The cognitive empathy scales assess the ability to imagine another person's perspective and the tendency to imagine oneself in the place of fictional characters. The emotional empathy scales assess the tendency to be concerned for the emotions of others and the emotional response when watching someone else feeling strong emotions. Mirella also tested the social skills of the children with the Interpersonal Competence Scale, a questionnaire completed by the parents. This scale measures how "popular" the child is, how many friends and playdates the child has, and so on.[5]

The first question Mirella wanted to answer was whether the brain of the typical ten-year-old shows the same pattern of activity as the adult brain, in which observation of other people expressing facial emotions activates three key neural systems: first, mirror neuron areas that provide an inner imitation (or simulation) of the observed facial expression; then the insula that connects mirror neuron areas with the emotional brain centers in the limbic system; and finally, the limbic system itself. During *imitation* of facial expressions, the same brain circuitry is activated, but even more strongly, since imitation "adds up" the neural activity of observation and action, as we have seen. Mirella's kids demonstrated exactly the same pattern of brain activity previously observed in adults. No surprise.

Her second and most important question was whether the activity in the mirror neuron system of the children could tell us something about their ability to empathize with other people and to have a successful social life. To find these answers,

Mirella correlated the scores obtained from the behavioral scales of empathy and interpersonal competence with the brain activity measured with fMRI. What she found is compelling. The emotional empathy scores of the children strongly correlated with the activity in mirror neuron areas during *observation* of facial emotional expressions. The more the child was emotionally empathic, the more the mirror neuron areas would fire up while the child watched other people expressing their emotions. Furthermore, the interpersonal competence of the children was also strongly correlated with the activity in mirror neuron areas during *imitation* of facial emotional expressions. Kids who were reported to be socially competent—had lots of friends and playdates—also demonstrated mirror neuron areas that would activate strongly during imitation.[6]

Studying these results, Mirella realized how critical mirror neurons are for social behavior. By simply looking at the activity of mirror neurons, an investigator can have a good grasp of the social abilities of the subject. It is as if the activity of mirror neurons is some kind of biomarker of human social competence—and a very nuanced biomarker at that, because emotional empathy, the ability to resonate emotionally with others, is mostly a private experience. It makes sense that Mirella found the correlation between emotional empathy and brain activity while children were simply observing other people's emotions. Likewise the tendency to overtly *mimic* the emotional expressions of others is an important component of successful social interactions. If you display intense happiness or deep sorrow and others respond with a stone face, you feel misunderstood. Reciprocal mimicry, as we have seen in chapter 4, is a key aspect of social interaction. Thus it makes sense that Mirella also found the correlation between mirror neuron activity during overt imitation of emotional facial expressions and interpersonal competence.

Her results with these children show us how the biology of the mirror neuron system is a cornerstone in shaping our empathic skills and our interpersonal competence early in life. But what if the development of the mirror neuron system is somehow altered or disrupted?

IMITATION AND AUTISM

Scientific reports on imitation deficits in children with autism date at least to the 1950s.[7] However, for decades such deficits were not considered important for understanding the fundamental cause of autism, mainly because the "theory theory" held sway, at first implicitly, then explicitly in the 1980s. As I explained in chapter 2, this model claims that children understand the beliefs and desires of other people because they have an innate module in their brain that helps them construct theories about other people, as if kids were little scientists testing their hypotheses on others.[8] Under theory theory, dysfunction in the hypothesized module in the brain would result in the sort of deficits demonstrated by children with autism, who generally fail what we call the false belief test. In this test, children are shown a little story involving Anne and

Sally. These children are typically impersonated either by pup-
pets or by real people. Sally and Anne are in the same room.
Sally puts her ball in a basket and covers it. When Sally
leaves the room, Anne moves Sally's ball from the basket to a
nearby box. At this point the subject children are asked:
Where will Sally look for her ball when she comes back? To
answer correctly, the child must realize that Sally did not see
Anne moving the ball and therefore holds the false belief that
the ball is still in the basket. The child who says that Sally
will look for the ball in the box clearly fails to see the situation from Sally's perspective. Such deficits were considered by
"theory theorists" the primary factor in the social behavior
problems of autistics. This idea reached its peak popularity in
the field in the early 1990s or so, even though it had one significant problem: autism was generally diagnosed between
two and three years of age (now even earlier!), whereas even
typically developing children do not consistently pass the
false belief test until they are about four. If two-year-olds with
and without autism cannot pass the false belief test, I would argue that it can hardly be considered a specific test for autism.

At the time—the early 1990s—imitation deficits were not
considered a primary deficit and were, accordingly, under-
studied. They were not, however, ignored by everyone. Swim-
ming against a strong tide, Sally Rogers and her colleague
Bruce Pennington at the University of Colorado Health Sciences Center suggested that imitation deficits in children
with autism had to be studied much more thoroughly, because
they could be the key to the social deficits of autism.[9] To this

day I am fascinated that they had such a great insight at a time when highly "mentalistic" (that is, cognitive) explanations were all the rage, and interest in their idea has increased, perhaps because the imitation deficits in autistics are quite visible. Imitation, however, takes many forms. Was imitation in children with autism impaired across the board? Peter Hobson at the University College, London, did not think so.

With his colleague Tony Lee, Hobson decided to test the hypothesis that children with autism imitate other people poorly because they cannot "identify" with those others. This hypothesis was inspired in turn by a series of previous studies Hobson had performed, all challenging the dominant view that the core problems of patients with autism are failures of a hyperrational theory of mind module. He argued that the main deficit was emotional. To make his case, Hobson and his colleague Jane Weeks had devised a very simple experiment to test whether children with autism and typically developing children would notice the same sorts of things about people.[10] To this end, they showed the kids pictures of either women or men, wearing either woolen caps or floppy hats, and showing either happy faces or gloomy faces. Weeks and Hobson asked the children to pick *one way* in which the pictures differed and to sort them accordingly in two boxes. Obviously, the children could have differentiated on the basis of gender, hat, or facial expression. In the first "round," both the typically developing children and those with autism used gender to sort the pictures. At this point, Weeks and Hobson asked them to sort the pictures again, without looking at gender. Here came the difference, and I'm sure you already know what that difference was: typically developing children picked the facial emotion as the sorting factor, whereas children with autism picked the hat. In his charming book *The Cradle of Thought*, Hobson suggests that it is as if "children with autism are almost blind to the feelings of others," as if they "are not moved by people's feelings."[11] Such results only encouraged Hobson in his conviction that cognition and "theory" deficits are not the problem for autistic children. The problem is the missing emotional connection.

Testing whether the imitation deficits of children with autism were also due to the children's inability to "resonate" emotionally with other people, Hobson and Lee came up with an experiment in which children could imitate both how people accomplished a goal and the "style" with which they conducted themselves—namely, in either a gentle or a harsh way. They tested typically developing children and children with autism, and the children were initially not even told that they were supposed to imitate what they watched. In a first demonstration session, Lee simply told all the kids, "Watch this." Then he performed simple actions with a number of objects. For example, he strummed a stick along a pipe rack or pressed a toy policeman, which then moved along under its own steam. When using the stick and pipe rack, he made a graceful and gentle strumming action for half of each group (the children with autism and the typically developing children), a harsh strumming for the other half. When he pressed the policeman with his hand, he used either his palm or two fingers. After that little game, for diversion's sake, the children were given a language test. Then Lee showed the children the pipe rack, the policeman, and other toys, and said simply, "Use these." What did the children do? It turns out that both typically developing children and children with autism used the toys to achieve the same goal that Tony Lee had demonstrated in front of them—for instance, strumming the stick along the pipe rack or pressing the policeman. However, while typically developing children also imitated the "style" Lee had adopted in front of them, children with autism did not. It seemed as if the children with autism were imitating the *action* that Tony Lee had demonstrated, whereas typically developing children were imitating the *person*, as Hobson puts it.[12]

The key point that emerges from Hobson and Lee's experiments and from other studies on imitation in children with autism is that the most critically impaired faculty is the social, affective form of imitation, more than the "cognitive" form of imitation (which, by the way, also shows some level of impairment). The critical impairment in patients with autism is social mirroring, which is supported by the neural interactions between mirror neurons and the limbic system through the insula. However, the data discussed so far on imitation deficits in autism is all behavioral. Is there more direct evidence of mirror neuron dysfunction in patients with autism?

THE MIRROR NEURON HYPOTHESIS OF AUTISM

A few years ago, two groups of scientists working independently suggested that autism may indeed be associated with a dysfunction in the mirror neuron system. One group in Scotland was led by Justin Williams, an expert in autism who teamed up with Andrew Whiten, an expert in imitative behavior in primates, and with Dave Perrett, an expert in monkey neurophysiology. The imitation deficits observed in children with autism, the neurophysiological properties of mirror neurons in monkeys, and the brain imaging experiments on imitation performed by my research group at UCLA led the Scottish team to hypothesize an early developmental failure of the mirror neuron system that would subsequently result in a cascade of developmental impairments leading to autism.[13] One of their key arguments was that imitation deficits may account for later deficits in "theory of mind," because both imitation and theory of mind require the child with autism to translate from the perspective of another individual to one's own perspective. If this is the case, they believed, the key neural deficit in autism was the dysfunction of mirror neurons.

At just about the same time, Vilayanur Ramachandran and his colleagues at the University of California, San Diego, were looking at the mu wave suppression of children with autism watching actions of other people. The EEG experiments described earlier in this chapter demonstrated that such mu wave suppression is considered a good index of mirror neuron activity. Like Justin Williams in Scotland, Ramachandran decided that mirror neuron dysfunction is a core deficit in autism and started testing his hypothesis empirically. The group in San Diego presented the preliminary data from their EEG experiments at the Society for Neuroscience meeting, the largest gathering of neuroscientists in the whole world, in November 2000.[14] This was the first presentation of evidence for the hypothesis of a mirror neuron dysfunction in autism, and the convergence of theoretical considerations and initial empirical findings inspired much more work on mirror neurons and autism. At least six different laboratories using a variety of techniques for studying the human brain have recently confirmed deficits in mirror neuron areas in individuals with autism.

BROKEN MIRRORING

Riitta Hari, who was responsible with Giacomo Rizzolatti for discovering that the suppressed mu rhythm during action and its observation is an important biomarker of mirror neuron activity, recently looked at how brain activity in patients with Asperger's syndrome, a milder form of autism, differs from "the norm." The patients were asked to imitate a series of simple movements of the mouth and face, such as protrusion of the lips, opening of the mouth, contraction of the cheek. Those movements are basically meaningless and do not express any particular emotional state.

I've described how MEG picks up the tiny magnetic fields surrounding the head that are created by the electrical activity of brain cells. It can detect events in the brain on the order of a few milliseconds. That's not much time at all, and it allows us to investigate the progression of activation in different brain areas *over time*. Taking advantage of this extremely high temporal resolution, Hari and her colleagues looked at the timing of activation in the mirror neuron system and found that patients with autism activated substantially the same brain areas as healthy volunteers during the imitations, but with *delayed* activation in the mirror neuron area of the frontal lobe (see figure 1, p. 62).[15] The neural communication between the mirror neurons in the parietal lobe and those in the frontal lobe was sluggish. The wiring was not working properly in these patients, creating problems in social behavior.

This experiment and others in different labs and different continents using different techniques—completed and published in scientific journals in the last few years—all converge in suggesting a mirror neuron deficit in patients with autism.[16] However, these studies were measuring activity in the mirror neuron system during tasks that had *no* emotional valence. Recall Peter Hobson's idea that individuals with autism have problems with imitation because they have problems in identifying with other people. A deeply felt mirroring that moves people closer to each other and makes emotional connectedness possible seems to be the main deficit of patients with autism. The other crucial issue that the raft of recent studies left unexplored was the functional significance of a mirror neuron deficit. No one had investigated whether the reduction of mirror neuron activity was at all correlated with the severity of the impairment of each patient with autism.

My wife, Mirella Dapretto, decided to investigate these

unexplored questions by building on her imaging experiment on typically developing children observing and imitating emotional facial expressions. In that experiment, she discovered that mirror neuron activity during such social mirroring tasks correlated with the social competence and empathic concern of the children. Mirella thought these same tasks would be ideal to test for social mirroring deficits in autistic children owing to mirror neuron dysfunction. Marian Sigman, a clinical psychologist at UCLA who has devoted her life to the study of autism, headed the group that assessed the severity of the autism in those children, all of whom were twelve years old. This assessment allowed Mirella to correlate the activity in the brain as measured by fMRI with the impairment of the children, and to test whether activity in specific brain areas is a faithful marker of the disease.

The results were right in line with all our predictions. While observing and imitating facial expressions, the children with autism had much lower activity in mirror neuron areas compared with typically developing children. Moreover, Mirella found a clear correlation between activity in mirror neuron areas and severity of disease: the more severe the impairment, the lower the activity in mirror neuron areas.[17] With this study, Mirella demonstrated that a deficit of mirror neurons is indeed a key factor for the social disorders of individuals with autism. And as every good scientific experiment will, Mirella's study also raised new questions, two urgent ones in particular: *Why* are mirror neurons dysfunctional in some children, and what can we do about it?

FIXING THE BROKEN MIRRORS

My hypothesis as to why the mirror neuron system fails to function in some children is inspired by the work of Ami Klin, a Brazilian psychiatrist working at the Yale Child Study Center. As a Brazilian, Ami is naturally a soccer fanatic, and when we meet, our conversation often drifts from the beautiful brain to the beautiful game. An endless point of discussion, obviously, are the matches between our two homelands in various editions of the World Cup. One day, Ami and I agreed that the Brazilian squad Italy managed to beat in 1982 was the best soccer team in the history of the game. It is still a mystery—to me, to Ami, to everyone—how the Brazilians lost that game. They seemed invincible. But they did lose, and Italy proceeded to win the Cup. One of the fascinating aspects of soccer is its unpredictability, which I guess reminds both Ami and me of the unpredictability of the human brain and of our research into its fathomless depths. When I discuss the brain with Ami, I love to go back to his work on how children with autism look at social scenes in a completely different way from typically developing kids. In those experiments, Ami and his colleagues at Yale used an eye tracker to monitor the "visual fixation patterns" of subjects with and without autism while looking at complex, dynamic social stimuli such as movie clips of people engaged in animated conversation. The subjects with autism do not look at the eyes nearly as much as the control group, and the greater the impairment in a subject, the greater the visual fixation on objects. In contrast, the better the social adjustment, the more the subjects with autism looked at the mouths of the observed people—not eyes, but mouths.

Another of Ami's fascinating findings is related to the striking dissociation in the performance of subjects with autism in spontaneous situations compared with well-structured ones. Using the invaluable eye tracker, Ami looked at how his two groups of subjects, those with and without autism, respond to pointing gestures. The observed scene was from the movie *Who's Afraid of Virginia Woolf?*, in which a character points at a painting on the wall and asks, "Who did the painting?" There are a bunch of paintings on the wall, so the verbal question alone is not very informative. However, a pointing gesture accompanying the question makes clear which painting is in question. Ami's eye tracker revealed that subjects without autism immediately and automatically follow with their gaze the pointing gesture and identify the designated painting. Subjects with autism, by contrast, do not follow the pointing gesture and move their eyes only after the verbal question is completed. They have no idea which painting the character is referring to and pointing at, so they shift their attention from painting to painting randomly. Later, the subjects were explicitly asked about the pointing gesture and its meaning. In this structured situation, they were able to provide an adequate answer with regard to the meaning of the pointing gesture in the scene they had just observed. However, during the spontaneous observation, their eye movements demonstrated that they had not grasped the meaning of what was going on between the two characters.[18]

Subjects with autism show these differences in visual fixations at a very early age. It is hard to say why this is. It is possible that they are less endowed with mirror neurons than other children and thus find watching the actions of others less rewarding. It is also possible that their differential pattern of visual fixation is not initially related to mirror neurons. Even in this case, however, there would be an effect on mirror neurons. As I described in chapter 4, a very likely scenario on how experience shapes and reinforces mirror neurons is that the reciprocal imitation during infancy allows the baby to make an association between certain kinds of movements and the sight of someone else making exactly those movements. Children who will develop autism tend not to look at the mother, the father, or the caregiver, and they cannot make the associations between their own movements and the movements of people imitating them. It follows that their mirror neurons cannot be shaped or reinforced. I believe this is a probable developmental scenario, which takes into account the properties of mirror neurons, the data from Ami Klin's eye-tracking work, and the role of imitation in early social interactions. Now all of us want to know whether these data and hypotheses could inspire forms of effective treatment to "restore" some mirror neuron functions in patients with autism.

I have been giving the same answer for some years. I believe that forms of treatment based on imitation may be very effective in helping patients with autism with their social problems. Right now, at least three scientists are studying the effects of imitation on children with autism: Jacqueline Nadel in Paris, Sally Rogers at the University of California at Davis, and Brooke Ingersoll in Oregon. I have recently watched one of Rogers's videos demonstrating the kind of intervention her group is doing with very young children with autism. When the child seems unengaged with others, Sally starts imitating the child, interacting playfully and emotionally. *Immediately*, the child is much more responsive to Sally and engages in emotionally meaningful interactions. How could such intervention *not* be beneficial?

Talking about Sally's video reminds me of an episode at a meeting organized by Cure Autism Now (www.cureautism now.org) in 2001. I had just finished delivering my lecture on mirror neurons, imitation, and the possible deficits in mirror neuron functions in autism. I had answered all the questions and left the stage when a man who works with patients with autism approached and said, "You know, what you are saying on imitation as a possible form of treatment makes a lot of sense to me. I work with severely impaired patients, and some days it seems really impossible to connect with them in any way. However, when everything else has failed, I have a last strategy that generally works quite well. Most of my patients make repetitive, stereotyped movements. When I do not know what else to do to establish a connection, I imitate the stereotyped movements. Almost immediately they see me, we finally interact, and I can start working with my patients."

We have seen how people tend to imitate each other, to

synchronize their movements, and how such synchronic motor behavior generally fosters a social connection. What is this immediate connection that imitation evokes? Although there is no well-controlled data on these spontaneous forms of imitation, it is likely that mirror neurons are involved. When the therapist imitates his patients, he may activate their mirror neurons, which in turn may help the patients to *see* the therapist, literally. This is only a theory of mine, but what we know about mirror neurons gives it some plausibility. Some time ago, Jacqueline Nadel in Paris sent me an extraordinary video of a twelve-year-old boy with autism, quite withdrawn, who shows a behavior often associated with autism: he makes stereotyped motor mannerisms with his hands (these motor mannerisms may take many forms; in this case, it is hand flapping). He is in a hospital room, alone, but with plenty of toys and objects to play with. In fact, there are two copies of every toy and every object. Another child comes in, a low-IQ girl without autism whom the boy knows well. She starts playing with some of the objects in the room and basically incites the boy to do the same. She puts on a cowboy hat, and then she puts the second cowboy hat on the boy. She helps the boy to put on one pair of sunglasses; then she puts on the second pair. The children shake hands and laugh. The stereotyped gestures of the boy with autism are rapidly disappearing. The girl now takes an umbrella, opens it, and parades around the room. The boy with autism spontaneously imitates her. Stereotyped gestures are *entirely* gone: he is a child fully engaged with his peer. The children play various imitative games for a while, sometimes the boy imitating her, sometimes the girl imitating him. When she leaves the room at one point, the boy almost immediately withdraws and starts his hand flapping again. When the girl comes back, those gestures disappear. The effect seems almost magical. Obviously it is not. Social mirroring connects individuals emotionally and may be a highly effective way of helping children with autism to overcome some of their social problems.

To test this hypothesis more rigorously, Nadel performed two experiments with autistic children. In both experiments, one group of children interacted with an adult who imitated the children, the other group interacted with an adult who simply played with them. Nadel found that the children who had been imitated by the adult demonstrated much more "social behavior" and engaged in much more reciprocal play with the adult than the children who had not been imitated. Furthermore, the children who had been imitated spent more time close to, beside, and touching the adult, compared with the children who had not been imitated.[19]

These extremely exciting results make a lot of sense in light of what we know about mirror neurons. In Oregon, Brooke Ingersoll has also used imitation as a treatment in children with autism, obtaining still more exciting results with naturalistic behavioral treatments. During seemingly spontaneous and playful interactions, the therapist starts to imitate the child's gestures, vocalizations, and actions directed at toys. Then the therapist invites the child to imitate her own behavior. Children exposed to this kind of treatment during naturalistic interactions show clear benefits, and these benefits go well beyond imitation alone. This is the important news.

Other social-communicative behaviors, such as language and pretend play, also show robust improvement. The techniques designed by Ingersoll can also be taught to parents, who can use them at home while spontaneously interacting with their kids. They can only help, and they do help.[20]

These techniques do not require any special technology and can be easily taught. They could be disseminated within the community of parents of children with autism quite rapidly and reach a large number of affected children. An awareness of the relationship between mirror neurons and imitation may promise potentially life-changing benefits for these children.

SEVEN

Super Mirrors and the Wired Brain

DARK WAVES IN THE BRAIN

In the spring of 2001, Vittorio Gallese and I, along with some other neuroscientists, were in Seville enjoying the spectacular local tapas. We got into a discussion of how the grass always seems greener on the other side of the fence. Vittorio, a great single-cell researcher, admitted that he envies the relative ease with which we brain imagers can perform our experiments, compared with the extensive training required for his monkeys, the surgery to implant the electrodes, and so on. I had to agree that the logistics of imaging are easier, but what about the results of our experiments? They require complex statistical analyses and even then are usually not so unequivocal as single-cell recordings. This can be frustrating. My point clearly resonated with my friend. He nodded his head. His eyes betrayed his excitement as he said, "When you find a beautiful neuron, it really is beautiful." Indeed.

We neuroscientists face tremendous barriers in our line of work. The type of research that allowed Vittorio and the team in Parma to discover mirror neurons in the first place by peering inside the brains of monkeys at the most exquisite, fine-grained level—the single cell—is invasive, requiring surgery to implant the electrodes. Although extreme care is taken by monkey neurophysiologists to avoid discomfort in the implanted subjects, the issues preclude such research on apes and humans (with rare exceptions, as we have seen, and with one more—the most important—still to come). Meanwhile, the incredible technology with which labs such as mine at UCLA study the human brain—fMRI, primarily—measures ensemble activity, that is, the activity of a large number of cells, and it has not been suitable for use with animals or even children, who are not very dependable about lying quietly without moving while inside loud machines.

Simply put, the various technologies lend themselves to different kinds of investigations, and each one is constrained by a unique set of factors, some practical, some logistical, some financial, and some ethical. With the monkeys, we haven't been easily able to move from the single cell to the ensemble, and with humans we haven't been easily able to move from the ensemble to the single cell. In this field, we have truly been caught betwixt and between. The conundrums have left us with inference as our main means of knocking down the fence and putting everything together, and inference, while valuable and necessary, is certainly not a perfect tool. It wouldn't be perfect even when working with our closest living relatives, chimpanzees; and the macaque monkeys are several evolutionary steps below both chimps and us. Unfortunately, there is little we can do to fill this gap. We cannot change the evolutionary process, and we definitely are not going to change our minds about the considerations that prevent the most invasive scientific investigations on humans and great apes. The society that does change its mind about this question is not one I would choose to live in.

We can't use single-cell techniques on humans, but what about using ensemble techniques on monkeys? Success here would allow us to compare single-cell and ensemble neural activity in the monkeys, and then to compare ensemble neural activity in monkeys and humans. These two sequential comparisons would certainly alleviate the inferential burden and help us put it all together. In recent years a number of researchers and laboratories have developed techniques to study the monkey brain with fMRI. Chief among them is Nikos Logothetis, at the Max Planck Institute in Tübingen, Germany. With an ingenious series of modifications of the standard hardware adapted for brain experiments with monkeys, Logothetis was able to perform *simultaneously* both single-cell recordings with intracranial electrodes and fMRI. This feat is impossible with the standard setup. Logothetis uses a modified setup in which electrodes do not burn the tissue around them (which standard electrodes would inside a scanner), do not pick up "artifacts" from the scanner activity, and do not create artifacts for the scanner's own signal. It is an amazing feat.

With his implanted electrodes, Logothetis measured the neural firing of individual cells while also measuring the changes in the fMRI signal. By doing so, he could test whether the firing of single cells is correlated with the brain signal measured by fMRI. Presented with visual stimuli, the visual cortex of the monkey's brain responded with increased discharge from individual cells and increased fMRI signal. This was a convincing correlation and real progress in the quest to align single-cell and ensemble neural activity. The progress can only pick up pace, because the number of laboratories using fMRI in monkeys will only increase.[1]

Then there's a third way to fill the gap between the single-cell recordings in monkeys and the imaging of ensemble neural activity in humans. I call this *opportunistic science*, which is already playing a significant role in neuroscience and, I believe, will be important for mirror neuron research. Before I tell you where we stand in this regard now and where I believe the future lies, I will show you how good it can get when opportunistic science combines with serendipity and answers one of the longest-standing and most important questions in the field of neurology.

This story takes us back to the era of positron-emission tomography (PET), which was widely used in the 1980s and early '90s but has been largely replaced by fMRI because it requires the use of radioactive material. As noted in chapter 2, one of the first imaging studies with the mirror neuron system was a "grasping" experiment with PET performed by Rizzolatti and colleagues. In those years, I was using PET to investigate brain regions important to our recognition of everyday objects—my first experiment with this technology. As required by the PET technique, we injected a small amount of radioactive material in the blood of the subject, who was asked to watch a variety of visual stimuli from a computer monitor. Simply put, the radioactive compound binds with certain molecules in the blood. To measure these molecules is therefore to measure overall blood flow. In the healthy brain, blood flow correlates with neural activity, and PET scanners measure brain activity by measuring blood flow. To ensure that our subjects received only a small amount of radioactivity, we had a limit of twelve injections of radioactive material—hence twelve brain scans—per subject, and we had to wait approximately fifteen minutes between injections to allow the radioactivity from the prior injection to be completely washed out.

One evening, I was doing the experiment on the fourth or fifth subject enrolled in the study. This twenty-one-year-old right-handed woman—who had volunteered, of course—told me at some point that she had a headache. I asked her whether she wanted to stop. She said she could continue, since all she was doing was watching stuff from a monitor. When we finished, I asked her a few questions. She said that the pain had progressively worsened during the roughly three hours required for completing the twelve injections and scans. Her description of a throbbing or pounding pain sounded like migraine. She also said that just before it started, she had noticed that she had blurred vision for a little while. This "aura" is a classic sign of migraine headache. Migraines affect women more often than men, and she said that this was not the first such headache she'd had. Putting all this together, we referred her to the neurology clinic at UCLA, where the detailed neurological history of her pain problems indeed suggested that she was a migraine patient.

I made copious notes on what had happened and when. Analyzing the PET scan data itself, I first looked, as dictated by our usual statistical protocol, for blood flow changes associated with the different kinds of stimuli the subject was seeing. In all of the previous subjects in this PET experiment, a notable increase in blood flow was recorded in the inferior temporal cortex when the subjects were watching everyday objects, as opposed to the other kinds of visual stimuli. With the migraine sufferer, however, I found *no difference* in blood flow between the various experimental conditions—a striking contrast with the clear effects observed in my previous subjects.

I "blamed" the migraine, but I did not know how to test my hypothesis. Since this was my first PET experiment and I had no experience in unusual data analyses, I turned to my collaborator and mentor in neuroimaging, Roger Woods, a neurologist who had developed a variety of analytical methods for brain imaging. When I told Roger what had happened, he said, "Well, let's look at the converted raw data." A quick explanation about PET data: the raw data are basically numbers that correspond to the amount of radioactive events detected in the scanner. The *converted* raw data are generally of no value, little more than blurred brain images in shades of gray. Even an experienced brain mapper will typically see no difference from one image to another. A statistical analysis by computer software is required to finally reveal even substantial changes. So when Roger suggested looking at the converted raw data, I was a bit puzzled. I hadn't been able to see anything of note in the statistical analyses, which are much more powerful than our eyes. What could we possibly see in the converted raw data? And of course I was wrong. Roger concocted some sort of animation with the twelve converted scans I had collected from our subject with the migraine. Looking at the animation, anyone could have seen that after the sixth scan, areas in the back of her brain became much darker, indicating a much weaker radioactive signal to the scanner and much less blood flow in those areas. The decrease in blood flow had to be extremely significant in order to be so easily detectable by naked eyes. Indeed, quantitative analyses later demonstrated a reduction in blood flow of about 40 percent in the dark areas—the back of the migraine patient's brain. From the sixth to the twelfth and final scan, the "darkness" spread progressively forward, from the back to the front of the brain of our subject, getting larger and larger. It was visually impressive in the animation.

Roger and I immediately realized that our serendipitous observation had resolved—literally overnight—the debate about the pathophysiology of migraine that had occupied neurologists and their patients for half a century. There were two main camps in this debate. One assumed that migraine was primarily a vascular problem. For largely unknown reasons (but with a variety of hypotheses proposed), blood vessels became initially constricted during a migraine attack (hence the symptom of the aura), then dilated, producing pain. The other camp assumed that migraine was caused primarily by a neural dysfunction, specifically by a phenomenon first observed in laboratory rabbits, called spreading depression. This phenomenon, described for the first time in 1944, is a dramatic but luckily transient reduction in electrical activity in the cerebral cortex. This depression of electrical activity literally spreads to contiguous cortical areas.[2]

Both theories had strengths and weaknesses. When PET became available in the 1980s, a consensus believed that the technology was ideal for resolving this very important debate. In theory, this was the case, but practical limitations had made it largely impossible to scan patients at the onset of what is, after all, an unpredictable event. Migraines don't follow a schedule. The spontaneous attack of migraine in our subject while she was in the PET scanner at UCLA was one of those lucky accidents that no neurologist or neuroscientist could even dream of. Moreover, the data we had collected were unequivocal. The rate of advancement of the "spreading depression" described in the lab rabbit in 1944 was very similar to the one we observed in our "dark wave" in 1994. This wave did not follow the anatomical territory of the major brain vessels, ruling out the vascular hypothesis. Only the neural hypothesis of migraine was supported by our observations.

This was a splendid case of opportunistic medical science. Indeed, one the reviewers who judged our scientific paper[3] used exactly these words. There are, however, other ways of doing opportunistic medical science, a bit more planned and requiring less serendipity than our migraine observations.

One of them is exemplified by the amazing work of Itzhak Fried and his collaborators at the Department of Neurosurgery at the UCLA Medical School. Fried and his colleagues are able to measure single-cell activity in the human brain. This extraordinarily important work allows us to fill the gap between monkey and human mirror neuron research.

IN THE DEPTH OF THE HUMAN BRAIN

Epilepsy is a neurological disease that affects millions of people, approximately 1 percent of the population. Symptoms can take many forms, but of course the most widely known is convulsions. In most cases, the disease can be effectively controlled with drugs. In some patients, however, the drugs are not able to control the disease well. The life of these patients is severely affected by their disorder, and one of the therapeutic strategies that may be adopted in these difficult cases is brain surgery. If it is possible to remove the focus of the epileptic activity—that is, the brain tissue from which the seizures originate—it is possible to control the symptoms effectively. Obviously, this decision is not be taken lightly, and most of the time all sorts of pharmacological treatments are tried first. When surgery is deemed necessary, it is imperative to find out exactly where the focus of the epileptic activity is located. We now have a variety of noninvasive tools for this purpose, but they are inconclusive in some patients. In these cases, the last-resort approach for the surgeons is to implant electrodes in a variety of brain regions and monitor activity over several days, perhaps even one or two weeks. Obviously, the location of these intracranial electrodes is dictated exclusively by medical criteria, not by experimental curiosity, and the patient's permission is of course mandatory. It is almost always granted. Then it is possible to collect unique and extremely valuable information of human neuronal activity in a variety of conditions, and at the exquisite resolution of single cells.

Possible, that is, because Itzhak Fried and his colleagues at UCLA had the idea of modifying the basic electrodes used for these intracranial recordings. Typically, the intracranial probe used with epileptic patients cannot record action potentials from single cells. In order to help locate the origin of the epileptic seizures, it doesn't need to. Indeed, the electrode can localize the epileptic locus by simply measuring the electroencephalographic signal—that is, the slow electrical waves that represent the ensemble activity of lots of neurons working together. The Fried team changed the basic setup and implanted a bundle of eight microwires dangling from the tip of each electrode. The tip of the microwire is small and sensitive: it can actually record action potentials from single cells. With eight electrodes usually implanted in each patient, we thus have sixty-four microwires from which single-cell activity can be recorded. Alas, not all microwires end up recording activity from single cells. It all depends on where the tip of the microwire ends up in the brain, which is something the surgeon—even the most experienced one—cannot control. The surgeon can control only where to place the tip of the electrodes in the brain. If the tip of a microwire dangling from an electrode ends up close to a neuron, it records its activity. If it ends up far away from the neuron, it does not. In a typical case, twenty to forty microwires can record cellular activity, and often more than one cell, generally two or three, can be recorded from one microwire. So it is quite possible to record from fifty cells in one session, and conceivably to record from twice that many (but no one ever gets that lucky).

With this unique setup, Fried and his collaborators are in a position to investigate with unprecedented detail the responses of the human brain to a variety of stimuli and situations. One of the first researchers to take advantage of this breakthrough was Roy Mukamel, a postdoctoral fellow in my lab who conducted an ingenious transoceanic experiment of sorts that exploited Itzhak's electrodes to provide more confirmation of the tight relationship between single-cell and ensemble activity in humans. First Roy showed segments of *The Good, the Bad, and the Ugly*, the famous spaghetti western directed by Sergio Leone, to healthy fMRI volunteers at the Weizmann Institute in Israel, where Roy had been a graduate student. Then he flew to Los Angeles to show the same segments of the movie to two epileptic patients in which Itzhak Fried had implanted intracranial electrodes with the microwires. Subsequently, Roy used the activity measured in single cells of the auditory cortex (as the name implies, that area of the brain that responds to sounds) of the two epileptic patients to test for correlated activity—measured by fMRI—in the brain of the healthy volunteers in Israel. Considering that the neurological patients were resting in quiet rooms in the UCLA hospital and the healthy volunteers in Israel were lying inside a very noisy fMRI scanner, it seemed almost inevitable that Roy's experiment looking at brain responses in the auditory cortex would fail. Wouldn't the radically different noise levels make moot the "auditory" readings picked up while the subjects were watching the movie? Astonishingly, this was not the case. Roy found close correlation between the single-cell activity in the auditory cortex of the epileptic patients in Los Angeles and the activity in the auditory cortex of the healthy volunteers measured in the fMRI scanner in Israel.[4] This data reinforced the results recorded by Nikos Logothetis, demonstrating a direct link between single-cell activity and the ensemble activity measured by fMRI. Both of these researchers, however, studied primary *sensory* areas: the visual cortex in the case of Logothetis, the auditory cortex in Roy's case. Can we generalize from these findings to the more complex cortical areas of the human brain—for instance, the frontal lobe areas that contain mirror neurons, or the temporal lobe areas that are repositories of our memories? Some very recent data, correlating single-unit recordings and fMRI signals in the temporal lobe during a memory task, suggest that in the temporal lobe, single-cell activity and fMRI activity do *not* go together. When subjects were asked to remember people and places, the individual neurons fired for only one specific person or place, but the fMRI activity increased for a large number of remembered people or places. How is that possible? The answer is in a neural phenomenon for which neuroscientists have found several different (and somewhat amusing) terms: the "grandmother cell," "sparse coding," and even the "Jennifer Aniston cell."

THE JENNIFER ANISTON CELL

A widely known term in neuroscience is "grandmother cell."[5] With this term, we succinctly refer to the theory that the brain may use single neurons to represent, recognize, and identify familiar objects. In its extreme versions, this theory suggests a one-to-one mapping between cell and object, such that your maternal grandmother is coded by one single cell in your brain, and your paternal grandmother by a different cell. There is a major theoretical disadvantage with the grandmother cell theory, and it's not hard to understand: if for some unfortunate reason a grandmother cell dies, your relationship with the recognized object—grandma—is completely destroyed! You are unable to recognize her or to remember her. On the other hand, there are also theoretical advantages with grandmother cells. The main one has to do with memory. If you need only one cell to retrieve the memory of your grandmother, you can memorize lots of different things with the multitude of cells in your brain.

In its most radical form, the grandmother cell theory is almost a caricature of a concept. The current version of this concept is labeled sparse coding. As the name suggests, this idea holds that a few neurons are selectively activated by a given stimulus—for instance, one's grandmother. Thus, the entire responsibility for remembering Grandma does not rest on a single cell, a much more efficient way to code for familiar objects and people. The death of one cell out of a small population coding the same stimulus would not be catastrophic.

One of the first descriptions of mirror neuron properties actually hinted at some level of sparse coding in the newly discovered system. In fact, the paper by Vittorio Gallese and the collaborators in Parma, published in *Brain* in 1996,[6] reports a variety of cells with selective properties to specific actions. Although the majority of the described mirror neuron cells responded to observed and performed grasping actions, almost half selectively responded only to specific actions such as placing objects, manipulating objects, holding objects, hand interactions, and so on. However, the empirical evidence that most robustly supports sparse coding, and indeed seems to support quite convincingly the idea of a grandmother cell, is the evidence reported recently by Itzhak Fried and his colleagues working with the intracranial electrodes implanted in the brains of the epileptic patients. The researchers used a laptop computer to present to the patients a large number of pictures of celebrities, famous buildings, objects, and animals. What they discovered was stunning. One cell responded only to various pictures of Bill Clinton, another only to pictures of the Beatles, another only to Michael Jordan, another only to the Simpsons cartoons. The Jennifer Aniston cell, as it was immediately dubbed, responded to several different pictures of the actress, but not at all to a large number of other stimuli, some of which were visually very similar to the pictures.[7] For instance, Julia Roberts triggered no response in the Jennifer Aniston cell. Amazingly, neither did a picture of Jennifer and Brad Pitt. Considering that this test was done when the two actors were still together and all over the tabloids and television, we have to wonder whether this sage cell was predicting their imminent *scandalo*, as we say in Italy!

There's more. Fried and colleagues also found a cell responding specifically to Halle Berry *and* to her name on the computer monitor. Such responses suggest that these cells may actually be coding more for memory than for vision. Indeed, the Jennifer Aniston cell responded to pictures of Jennifer Aniston and Lisa Kudrow, which seemed to associate the two actresses because of their roles in the same sitcom, *Friends*.[8] The fascinating findings continue to pour out of this research with the epileptic patients. The UCLA group has been able to document single-cell discharges during visual imagery (when you just think about seeing places and people) and during memory for places.[9] Cells that respond to certain types of visual stimuli—say, to a face—also discharge when the patient is simply imagining the face.

As it happens, Itzhak and I had collaborated in the mid-1990s in a study on how visual and motor information gets integrated from the left brain to the right brain and vice versa through the corpus callosum, a large and flat bundle of hundreds of millions of axons—the neurons' long extensions—connecting the left and right brain. This was before Itzhak started recording with depth electrodes in epileptic patients and before I became interested in mirror neurons. Obviously, I knew about his new work and he knew about mine, and in retrospect it seems almost inevitable that at some point we would collaborate again in an attempt to record individual mirror neurons, but we put the prospect aside for a couple of years when it first came up. There seemed to be a fatal mismatch of requirements. My idea was to look at the areas in the human brain we know should contain mirror neurons, because they are the human homologues of the monkey mirror neuron areas. Unfortunately for me, Itzhak typically implanted his electrodes in the restricted set of locations that tend to be more "epileptogenic" than others—that is, more likely to be the source of the seizures. Our needs were at loggerheads: his mandated locations were not my required locations, and he practically never implanted electrodes where I thought I needed them. The collaboration languished, and when that happens in science, it is often hard to get it going again. Scientists are so often swamped with their own research and ongoing collaborations that new ones that may seem natural and obvious never get going without some kind of trigger or lucky break.

In our case, that good fortune came in the person of Arne Ekstrom, a postdoctoral fellow working with Itzhak on cells in the human brain that are important for memory for places. Arne had just designed a task in which patients drive a taxi in a city they do not know—a virtual reality environment, of course—and he noticed that while one of his patients pressed a key to perform some aspects of the task, certain cells in the frontal lobe had very high activity. They definitely seemed to be motor cells. Arne also knew about my work on mirror neurons and wondered whether we could test mirror neuron properties in the frontal cells recorded by Itzhak in the epileptic patients. Furthermore, Arne was interested in the gap between single-cell research with intracranial electrodes and the ensemble results from brain imaging experiments using fMRI.

In discussing this potential project with him, I suddenly realized how I could use Itzhak's electrodes after all.

IN SEARCH OF SUPER MIRROR NEURONS

If mirror neurons are such powerful neural elements that help us reenact in our own brains what other people do, as I believe they are, the evolutionary process that created such a neural mechanism must also have created some form of control over it. It would be highly inefficient for us, after all, to imitate observed actions all the time. Moreover, imitation takes many forms, and some of these forms are highly complex. Ap Dijksterhuis, a social psychologist from Holland, draws a distinction between the complex forms of imitation he calls the imitation "high road," as opposed to the "low road" of straightforward motor imitation (grasping the cup, for example). Regarding his high road, Dijksterhuis gathered an impressive array of behavioral data confirming various forms of complex mimicry in human behavior. Let me give you just one fascinating example. In a series of experiments, one group of participants was asked to think about college professors, who are typically associated with intelligence, and write down everything that came to mind, and a second group was asked to do the same regarding soccer hooligans—those unruly and destructive fans who are typically associated with stupidity (or at least with very stupid behavior). Then both groups were asked a series of "general knowledge" questions, a task ostensibly unrelated to the first one. But it turned out

that there was a relationship: the participants who had con-
centrated earlier on college professors *outperformed* the participants who had been thinking about soccer hooligans. Indeed,
the "college professor" participants outperformed a control
group that came fresh to the general knowledge questions, and
this control group in turn outperformed the "soccer hooligan"
participants.

Conclusion: just thinking about college professors makes
you smarter, whereas thinking about soccer hooligans makes
you dumber! Ap Dijksterhuis summarizes his research by
saying, "Relevant research has shown by now that imitation
can make us slow, fast, smart, stupid, good at math, bad at
math, helpful, rude, polite, long-winded, hostile, aggressive,
cooperative, competitive, conforming, nonconforming, conservative, forgetful, careful, careless, neat, and sloppy."[10]
That's quite a list, and I believe that this constant automatic
mimicry is indeed an expression of some form of neural mirroring. At the same time, such high-road imitation often involves a series of fairly complex and subtle behaviors, and it is
hard to believe they can be implemented by the "monkey see,
monkey do" cells discovered in Parma. Even though some
mirror neurons in monkeys do show a more sophisticated
form of firing—recall the "logically related" cells that fire not
for the same observed and executed action, but for logically
related ones, such as placing food on the table and grasping it
and bringing it to the mouth—one has the feeling that even
this pattern is hardly sufficient for imitating complex aspects
of human behavior.

I decided that the subtle imitation of complex behavior we

humans do all the time very likely requires a broader concept of the mirror neuron system, one that encompasses cells whose role is control and modulation of more classical and simpler mirror neurons. This higher order of mirror neurons may be called super mirror neurons, not because they have superpowers, but because they may be conceptualized as a functional neuronal layer "on top of" the classical mirror neurons, controlling and modulating their activity. After developing these initial ideas on super mirror neurons, I asked, as any diehard brain mapper would ask, where these super mirror neurons might be in the brain. I came up with three brain regions that are located in the frontal lobe (the front of the brain) and are connected with the frontal mirror neuron area. The names of these brain regions are: the orbitofrontal cortex, the anterior cingulate cortex, and the presupplementary motor area. After talking with Arne Ekstrom about his suggestion, I realized that all of these regions *are* in the frontal lobe areas in which Itzhak Fried implants his electrodes. I became aware that we could go after not the classical mirror neurons investigated in the monkey, but rather these hypothetical super mirror neurons. The Fried-Iacoboni collaboration was back on track! We have so far obtained recordings for approximately sixty single cells that show mirror neuron properties in the frontal lobe areas hypothesized for super mirror neuron activity. Some of these cells have a very interesting pattern of neuronal firing. The firing rate increases while the patient performs the action, as in monkeys. In sharp contrast with mirror neurons in monkeys, however, these cells shut down entirely while the patient observes the action.[11] This pattern of activity suggests that these cells may have an inhibitory role during action observation. By shutting themselves down, they may tell the more classical mirror neurons, and other motor neurons as well, that the observed action should not be imitated. Furthermore, this differential coding for action of the self (increased firing rate) and for actions of others (decreased firing rate) may represent a wonderfully simple neural distinction between self and other implemented by these special types of super mirror neurons. At the end of chapter 5, I proposed that mirror neurons help carve out a proper sense of self and other from the primary intersubjective sense of *us*. This process is most likely implemented by these special super mirror neurons. Indeed, the brain areas from which we recorded these cells are the least developed in early infancy and show the most dramatic developmental changes later on.

The presupplementary motor area is also known to be important for putting simple actions together into more complex motor sequences. The recordings of mirroring responses in some of these cells especially intrigue me. Mirror neurons in this region (or, better, super mirror neurons in this region) would seem to be the ideal brain cells for organizing simpler imitative actions—the low road—into complex forms of imitative behavior—the high road.

Unfortunately, not all the complex forms of imitative behavior are good for you or for us (society at large). It is time to explore mirroring as a social phenomenon that may induce what is called in scientific terms "problem behavior."

EIGHT

The Bad and the Ugly:
Violence and Drug Abuse

THE BAD: THE CONTROVERSY ABOUT
MEDIA VIOLENCE[1]

In the spring of 2002, a fourteen-year-old student attending a private Catholic school in France was tortured by two classmates who considered her "too pretty." The knife they used resembled the one used in the movie *Scream*, which the older of the two torturers had apparently just seen. These two girls came from stable middle-class families, studied at an elite school, and had never been accused of any act of violence. Then, a couple of weeks later and on another continent, two teenage boys went on trial in North Haverhill, New Hampshire, for the murders of college professors Half and Susan Zantop. The Zantops were killed in their home, by multiple stabbings, apparently during an attempted robbery. During the legal proceedings, it emerged that the boys owned and enjoyed playing with an interactive and especially realistic video game in which the player stabs his victims and watches them bleed to death.[2]

Were these two crimes instances of "imitative violence" induced by media violence? Obviously, nobody can have a definitive answer. Cause-and-effect relationships can be complex, and we now have a large literature, featuring a wide variety of opinions, on this question. These studies fall into three major categories. One features experimental manipulation of exposure to violence. The strength of these studies is that they are very well controlled and can assess the effects of exposure to violence on subsequent manifestations of violence in a fairly accurate way. Their weakness is that the artificial laboratory settings cannot fully replicate the complexity of real-life situations. A second category is correlational studies, in which violent behavior and exposure to media violence are measured in a large number of individuals. Their strength is that they look for relationships between media violence and violent behavior in real life. Their weakness is that they cannot fully demonstrate whether the relationship between exposure and behavior is due to the causal influence of the media or to the inclination of inherently violence-prone individuals to watch violent media and to play violent video games. The third type of study tries to address the problems encountered in the other two by repeatedly measuring over a long period of time violent behavior and exposure to media violence in large groups of subjects, typically hundreds. These studies can therefore determine whether the exposure to media violence actually precedes violent behavior.

Perhaps the best way to reach some conclusions on this complex and important issue is to integrate the information from these studies and to look for any consistency in their findings. In doing so now, I will address the question most pertinent for this book: Could mirror neurons be part of the problem?

The results of the controlled experiments with children in laboratory settings could not be more clear and unequivocal: exposure to media violence has a strong effect on imitative violence. Typically, these experiments are performed by showing children a short movie. Some of the movies are violent; others are not. The children are then observed while interacting with one another or while playing with objects like Bobo dolls, which spring up after being knocked down. A consistent finding is typically observed in these experiments. The children who watched the violent short movies display much more aggressive subsequent behavior toward both people and objects than the children who watched nonviolent short movies. This effect of media violence on imitative violence is observed in children from preschool to adolescence, in both boys and girls, in both naturally aggressive and nonaggressive children, and in different races. The results are quite convincing.[3]

What is left unanswered by these studies is the real-life impact of media violence on people's (including older people's) actual conduct in the world. Are the effects demonstrated in the experimental setup long-lasting or transient? Artificial or real-world? The findings from the correlational studies suggest that causal relationship is long-lasting and real-world. For instance, a study that retrospectively analyzed the rash of bomb threats and other threats of school violence among Pennsylvania school districts following the Columbine High School massacre reported more than 350 threats of school violence in the fifty days after the Columbine massacre, in comparison with the one or two threats per year estimated by school administrators before the massacre. Furthermore, children who watch more media violence tend to be more aggressive than other children. These findings are highly reproducible across studies and even across countries.[4] Pairing the findings from the correlational studies with the findings from the laboratory experiments with children does tempt us to conclude that media violence inspires imitative violence, but the best empirical data must come from the longitudinal studies that investigate the correlation between watching media violence and violent behavior *over time*.

One of the very first such studies was initiated in New York State in the 1960s and included almost a thousand kids. Even while controlling initial aggressiveness and other major variables, such as education and social class, this study demonstrated that watching media violence in early childhood was correlated with aggressive and antisocial behavior approximately ten years later, after high school graduation. These findings are already impressive, but there is more: the same boys were followed for more than another decade, for a total of twenty-two years of follow-up since initial enrollment in the study, and again the results were clear-cut: both early viewing of media violence and early aggressive behavior correlated with criminality at age thirty!

A later study looked at the *differences across nations* in imitative violence induced by media violence. Even given the cultural differences and the differences in the style of the television shows in the countries in the study—the United States, Australia, Israel, Finland, and Poland—similar results were observed across nations with regard to early effects on imitative violence of watching media violence. However, there were some differences across nations, suggesting that the effects of the cultural environment can modulate the impact of media violence. A striking result was obtained within Israel, where the effects of watching media violence were observed in children living in cities, but not in children living in a kibbutz.[5]

A still more recent longitudinal study on American children has provided some of the most impressive empirical results in support of the hypothesis that media violence induces imitative violence. This was a fifteen-year follow-up study on children's exposure to media violence. Several kinds of assessments of violence and aggression were used to correlate the behavior of the subjects when they were twenty-one to twenty-five years old and their exposure to media violence when they were six to nine years old. The study found strong correlations, even after controlling for confounding factors such as individual aggressiveness, intelligence, poor parenting, and social class.

Taken together, the findings from laboratory studies, correlational studies, and longitudinal studies all support the hypothesis that media violence induces imitative violence. In fact, the statistical "effect size"—a measure of the strength of the relationship between two variables—for media violence and aggression far exceeds the effect sizes of passive smoking and lung cancer, or calcium intake and bone mass, or asbestos exposure and cancer.[6] *Still*, all of this impressive behavioral data tend to be looked at with some skepticism, if not outright hostility, often with the argument that correlation, no matter how strong, is not necessarily causation. This theoretical point is correct, of course—and it was also the argument used by the tobacco industry for the better part of the entire twentieth century to dispute the link between smoking and lung cancer. So the skepticism and hostility might be honest or they might be duplicitous, but they have been helped either way by the lack of neuroscience data revealing the underlying neurobiological mechanisms of imitation. Now that gap in our knowledge is shrinking rapidly, thanks to the discovery of mirror neurons. The implications of the discovery are far-reaching, not only for our understanding of imitative violence and possible decisions to address it, but even in philosophical terms. Many long-cherished notions about human autonomy are clearly threatened by the neuroscientific scrutiny of the biological roots of human behavior. Our notion of free will is fundamental to our worldview, yet the more we learn about mirror neurons, the more we realize that we are not rational, free-acting agents in the world. Mirror neurons in our brains produce automatic imitative influences of which we are often unaware and that limit our autonomy by means of powerful social influences. We humans are social animals, yet our sociality makes us social agents with limited autonomy. Should we deny this biological reality on the grounds that explaining the social influences that produce evil may eventually condone it? I believe it would be more logical to use our understanding of the biological roots of our limited social autonomy to *prevent* evil. To do so, we need to abandon the long-standing belief at the basis of the "argument from autonomy," which is the topic of the next section of this chapter.

ARE WE AUTONOMOUS AGENTS?
MIRROR NEURONS AND FREE WILL

Most discussions of imitative violence distinguish between the short-term effects of watching media violence and the long-term effects. Clearly, classical mirror neurons and super mirror neurons are plausibly involved with two of the short-term effects: immediate imitation of violent behavior and a general arousal owing to observing violence. We have already seen in several contexts the pervasive nature of human imitation and the critical role of mirror neurons in this imitation. The neural properties of these cells can easily explain the immediate imitation of violent behavior, especially simple acts of violence, just as they can explain, as we have seen, the mirroring of smiling, foot shaking, face rubbing, and so on. If the chameleon effect makes us imitate what we see, we also need some way to inhibit such imitative behavior. Otherwise we are in big trouble. As we have just seen in chapter 7, one of the major roles of super mirror neurons may be just such an inhibition of the more classical mirror neurons, such that when we see somebody else making an action, we do not compulsively imitate that action. The observation of violence is presumed to cause arousal. This arousal, in turn, may facilitate imitative violence by reducing the inhibitory activity of super mirror neurons, such that the imitation of violent behavior is less effectively inhibited. Although nobody has yet performed an experiment to test the neural mechanisms I am proposing here, they are plausible in light of what we know about the human mirror neuron system, and they could be tested in brain imaging experiments in the near future.

Consider now the long-term effects of media violence. Classically, these have been ascribed to complex forms of imitation in which individuals observing aggressive behavior not only acquire complex coordinated motor behaviors that make them aggressive and violent but also become convinced in the process—in an unconscious way—that such behavior is a good way of solving social problems. I have hypothesized that super mirror neurons could account for this complex imitation by providing the generalized capacity to put together simpler forms of actions in order to produce a complex, coordinated pattern of behavior. Thus both short-term and long-term types of imitative behavior induced by the observation of media violence seem to map fairly well onto the functions of mirror neurons and super mirror neurons. Previously we have seen how mirror neurons can undoubtedly be good for us, enabling our feelings and actions of empathy for others, but they also provide a compelling neurobiological mechanism underlying imitative violence induced by media violence. As I noted earlier, however, the evidence for this link has had a hard time gaining traction, and the argument based on the allegedly missing proof of causality provided good cover for all kinds of actual motivations, including the powerful financial interests of the media industry. Violence sells, so it is clearly convenient to deny a causal link between media violence and violent behavior. There is also the concern that the emphasis on media violence may lead to forms of censorship, a very serious question. Second Amendment purists also fear (probably with good reason) that this issue will leak into the debate over gun control in the United States. Finally, we are naturally inclined to think of ourselves as autonomous individuals who are not going to be influenced in any direct, slavish, monkey-see, monkey-do way by what *we* see. The data on imitative violence clearly threaten this precious notion.

Indeed, it has been argued that even with the evidence strongly linking media violence to imitative violence, this "argument from autonomy" prevents any form of intervention. Harmful "speech"—intended here in a very broad sense, including movies, television, and video games—is typically protected under the assumption that the effects of any kind of speech are always undergoing the mental intermediation of the listener or viewer. According to this reasoning, even individuals almost certainly "guilty" of imitative violence are nevertheless fully responsible for their own actions, while the media that distributed the imitated violence is not responsible at all. According to free-speech theorists, we are all rational, autonomous, and conscious decision makers. However, the data we have reviewed in this book—ranging from the unconscious forms of imitation observed while people interact socially to the neurobiological mechanisms of mirroring that have their key neural elements in mirror neurons—suggest a level of uncontrolled biological automaticity that may undermine the classical view of autonomous decision making that is at the basis of free will.[7]

The implications of these considerations are obviously important for every society. They bring to the foreground fundamental questions of ethics, justice within the legal system, and public policy. The questions raised by the discovery of mirror neurons force us to rethink, or at least to consider with new eyes, some of our fundamental assumptions. Indeed, a whole new discipline is emerging, named neuroethics.[8] Its meetings have such titles as "Our Brain and Us: Neuroethics, Responsibility, and the Self" (held at MIT in Boston in 2005).

The classical conflict between those who emphasize the biological determinism of human behavior and those who insist that our ideas and social behavior rise above our neurobiological makeup has never considered the possibility that our neurobiology dictates our social behavior to begin with. I believe that a better understanding of the neurobiological mechanisms shaping human social behavior—in particular the research on mirror neurons—should also directly inform the making of our social codes, and I will elaborate on this in the final chapter of the book. Our instinct for empathy is part of the good news stemming from mirror neurons. Imitative violence could well be the bad news—and there may be more. Another possible negative implication of mirror neurons on our behavior is their role in the wide variety of addictive behaviors and the subsequent relapses, to which we are so prone.

THE UGLY: ADDICTION AND ITS RELAPSE AFTER QUITTING

One of the major problems in the treatment of drug addiction is relapse. In a variety of studies of addictions to smoking, alcohol, and drugs, the rate of relapse after a period of abstinence varies from 30 percent to as much as 70 percent. These are very large numbers. What can be done to reduce them? On logical grounds, an essential first step would be identification of any markers that can tell us ahead of time which subjects are more likely to relapse. Such identification of at-risk subjects would allow more individualized and effective preventive measures to reduce relapse.

There is at least one such marker, and an obvious one it is: *craving* the drink, the smoke, the fix. Perhaps surprisingly, not all addicts experience, or at least they don't report, the same degree of craving. Not surprisingly, though, the greater the residual craving during the treatment for addiction, the more likely the relapse later on. In everyday life, craving may be induced by a variety of social cues, such as the cues picked up by an addicted smoker from other smokers. Among smokers attempting to quit or who have successfully quit for a period of time, interaction with other smokers, especially within the same social group, is one of the most significant predictors of relapses.[9]

In collaboration with Edythe London, a world-renowned figure in the neurobiology of drug addiction, and John Monterosso, another expert in the field, my lab at UCLA is currently exploring the hypothesis that mirror neurons may be responsible for relapses among smokers who have successfully quit the habit. Our analysis is as follows, and we believe it probably applies to all sorts of addictions. When former smokers watch other people smoking, their mirror neurons are automatically activated, because facilitating some kind of inner imitation of the actions of others is *what mirror neurons do*, as we have seen time and again. As a former smoker, you have used your hands to light a cigarette a million times, and you have done this when you wanted to smoke. The activation of your mirror neurons therefore also activates the association of the motor plans required to light the cigarette and bring it to the mouth. Rare indeed is the smoker who has just quit and remains unaffected by seeing another smoker light up and inhale that first delicious puff. According to our theory of mirror neurons, such obliviousness would be virtually impossible.

This analysis led us without pause to two hypotheses regarding the links between mirror neurons and smoking. One, mirror neuron areas of smokers are activated much more strongly by the sight of other smokers, compared with mirror neuron areas of nonsmokers. Two, the higher the activity in the mirror neuron areas of smokers, the more intense the craving to smoke. Neither of those hypotheses will surprise the reader. They follow necessarily from everything we have learned. In chapter 1, I introduced the experimental evidence suggesting that mirror neurons in monkeys are shaped by experience and can learn new properties. Initially, these cells did *not* to respond to the observation of actions that did *not* belong to the motor repertoire of the monkey, such as using a tool to pick up food. But then they did, and this new response of some mirror cells is probably due to the fact that this particular monkey has repeatedly seen humans using tools in the lab (if only in the persistent experimenters' attempts to elicit a discharge in the animal's mirror neurons).

There is also evidence that the human mirror neuron system is shaped by experience. In a couple of famous imaging experiments performed in London, fMRI was employed to measure the brain activity of two groups of dancers watching videos. In the first experiment, the scientists compared the brain activity of classical ballet dancers and capoeira experts as they watched performances of each art form. Capoeira is a Brazilian martial art featuring unbelievably acrobatic moves to the accompaniment of distinctly rhythmic music. The movements of capoeira and classical ballet are quite different, and the scientists in London had the clever idea to use this fact in their mirror neuron study. They found that classical ballet dancers had higher mirror neuron activity than capoeira experts while watching classical ballet videos, while the capoeira experts had the higher activity while watching capoeira videos. But were these differences truly due to the subjects' identification with the observed movements, or could they have been simply responses to cultural differences? A second study involving only classical ballet dancers addressed this question by utilizing the fact that male and female ballet dancers perform certain gender-specific movements. (Most obviously, only the men lift their partners, only the women dance on toe.) In the experiment, men and women ballet dancers watched videos of only those ballet movements performed by one or the other sex. The results reflected the results of the first study. The female ballet dancers had higher mirror neuron activity than the men while watching their distinctly female movements, and the men had the higher activity while watching their own movements.

A recent imaging experiment with nursing experts and novices watching nursing activities also found that the degree of expertise modulated brain activity in the observers. All these brain imaging data clearly show that experience shapes mirror neuron activity while we observe others performing actions.[10] On this basis alone, it makes sense to predict that smokers have greater mirror neuron activity compared with nonsmokers while they watch other people smoking or lighting up a cigarette, and our preliminary data support our prediction. For the experiment, we showed our subjects (all of whom smoke at least fifteen cigarettes per day) a series of ten-second video clips. In half of these, the observed individual is either opening a pack of cigarettes or smoking a cigarette (in various environments, indoors and outdoors). In the other half, the observed individual is opening a soda can or a bottle of water, or drinking soda or water. Of course, mirror neurons were activated by the observation of both types of actions, smoking and drinking, but mirror neuron areas had much higher activity during observation of smoking-related actions.

Only time will tell whether our second prediction—that is, the higher the activity in mirror neuron areas while watching other people smoking, the more intense the craving to smoke—will be supported by the empirical data. If the prediction is correct, we may obtain a potentially important biomarker for craving and, possibly, for the likelihood of relapse after quitting. Obtaining such a biomarker may help those professionals working with addicts to identify individuals who need a somewhat different form of treatment, more tailored to their individual needs. Alternatively, these individuals may be helped to prevent relapses by actively intervening in their environments to remove the social cues that are more likely to induce a relapse. Given the high rate of relapse for practically every known form of addiction, a better understanding of the role of mirror neurons in relapses will be extremely important in the treatment of addictive behavior. And if mirror neurons can potentially tell us whether an individual is more or less prone to fall back on his or her addiction, one wonders if they can tell us even more about how all of us make our decisions.

NINE

Mirroring Wanting and Liking

THE NEUROSCIENCE OF BUYING

Have you ever been in a focus group? Me neither, but I know how they work, and I know they are not reliable, for several reasons. First, people have the inclination to say what they believe the interviewer or moderator wants to hear, rather than what they really think. The social pressure to say the proper thing often overwhelms true opinions. Second, the dynamics of focus groups are inherently "noisy," and we now know about the power of imitation in human social interactions. When one alpha-type participant in a focus group is especially insistent in expressing his or her opinion, the others tend to tilt in that direction. It is human nature, and I am not alone in noting how this nature affects focus groups. Marketing professionals know well the pitfalls of the classic marketing tools that require one person to ask questions of another person, but what else do they have to work with? Until now, very little.

As it happens, I believe the problem for the pros is even worse than they imagine. Another powerful aspect of human behavior complicates the idea of relying on what people say in an attempt to understand their decision-making processes. Some readers may not appreciate where this discussion is now heading, because of course we want to think of ourselves as autonomous agents, in control of our own lives and able to make decisions in a rational and conscious way. However, there is plenty of evidence to suggest that this is not necessarily the case. A wealth of psychological studies show that our ability to re-present our own original experiences and decision-making processes may be limited.[1] Jonathan Schooler—on the basis of empirical evidence from several labs—proposes two levels of dissociation between consciously experienced mental processes and meta-conscious mental processes (that is, re-presented mental processes, as when subjects explain why they think that one ad is better than the others). He calls the first type temporal dissociation. A typical example is when we first zone out while reading and then catch our mind wandering and paying no attention to the written page. The sudden realization that our mind has wandered reveals a temporary lack of explicit awareness of zoning out.

Schooler calls the second type translational dissociation. This type of dissociation is more relevant to my argument. The main idea here is that information gets lost or distorted when subjects explain with words their own experience.

These translational dissociations are especially evident when subjects report experiences that are inherently difficult to put into words, such as faces, colors, cars, voices, wines, and so on. Studies have shown that detailed verbal descriptions of various experiences impair memory. This phenomenon is called verbal overshadowing. In these experiments, participants first watch a picture of a face and then are either asked to describe the face in detail or to engage in an ostensibly unrelated verbal task. In a subsequent recognition test of a different picture of the same face, the subjects that had described the face in detail performed worse than the other subjects.

There are several kinds of translational dissociation. Let me give you only two experimental examples: thinking aloud during problem solving has been proven to disrupt performance; being subliminally primed with the words "thirst" and "dry" has shown increased drinking, but at the same time it did not affect the subjects' self-reported thirst.

A recent study demonstrated the dissociation between verbal reports and perception in a dramatic way. Male subjects were asked to evaluate the attractiveness of two female faces, selecting the more attractive face on the basis of pictures on cards. Immediately after the subject made his choice, the experimenter removed the cards. A few seconds later, the experimenter showed the subject one of the two cards and asked him to explain why this face was more attractive. The tricky part of this experiment is that in some trials the experimenter showed the subject the *unselected* picture—that is, the picture of the woman deemed less attractive. You would think that the subjects who were shown the unselected picture would immediately figure out that they were being tricked. Amazingly, only about 10 percent of the manipulated trials were detected. One out of ten! We now call this phenomenon choice blindness. It seems that we humans are blind to our own choices. This evidence is clearly difficult to reconcile with the idea that we are rational decision makers in complete control of our own decisions. Almost embarrassingly, when subjects did not pick up on the trick, they proceeded to offer good reasons why this initially unselected face was the more attractive one. In fact, there was substantially no difference between the explanations provided for the truly chosen faces and for the switched ones.[2] Is it possible that subjects were aware of their mistakes and simply decided to remain silent because they were embarrassed? Unlikely. Indeed, the moment participants realized that they were tricked, they became highly suspicious of the whole experiment, and subsequent trials had to be removed from the analysis.

Given this evidence, how can our verbal reports on how we make our own decisions be entirely trusted? Enter neuromarketers, who advocate using neuroscience to better understand and predict human behavior. The time is ripe for the application of brain imaging to a variety of aspects of our society. Our knowledge of the neural mechanisms associated with behavior is increasing at a rapid pace. Brain scanners are becoming more readily available. Using them to look at brain activity, we can have a much better grasp on what's really going on as we make our decisions and decide what to buy.

For some reason, people who hear the word "neuromarketing" sometimes equate it with a sophisticated form of mind control. I am not sure what the logic behind this claim might be, but I do know that the claim itself is based on a misunderstanding. Mind control requires some form of manipulation. Neuromarketing does exactly the opposite. It reveals to consumers and marketers what people truly like. It even makes consumers more aware of their own deeper motives—motives that, as we have just seen in the cases of translational dissociations, consumers cannot explicitly verbalize.

The specific neural system that neuromarketers have typically focused on so far is the so-called reward system. With this term we neuroscientists describe a set of brain areas that are associated with reward-related behavior. Of course, this behavior is complex and has been subdivided into several components, including the incentive value of rewarding stimuli, the approach and consummatory behavior for acquiring rewards, the emotions associated with rewards, the "expectancy" of rewards, and so on.[3] Not all aspects of this complex system have been entirely mapped out, but a large amount of work now allows at least some of the major concepts emerging from this work to be applied to other fields, such as marketing.

In evolutionary terms, the reward system has probably evolved from a system that evaluated primary goals such as food and sex to one that evaluates much more culturally driven stimuli that have a rewarding quality for modern humans. Given this plausible hypothesis, observing the reward system in the brains of subjects in the fMRI scanner who are watching products and advertisements should be worthwhile for both academic researchers and marketers. Probably the first such brain imaging study to adopt this strategy involved male subjects looking at different kinds of cars. In this German study, twelve subjects who had rated themselves as "highly interested" in cars were presented with pictures of three categories: sports cars, limousines, and small cars. Controlling as much as possible for the variety of "low-level" visual aspects of the stimuli, the brain imagers used black-and-white photographs, all with the same orientation of the car. They even removed all brand names. Indeed, to judge by the sample pictures displayed in the paper that reported the findings, this controlling process for the photographs made the cars rather unattractive (at least to me, definitely not a car buff). At any rate, each picture was presented for six seconds, and the subjects were asked to rate the attractiveness of each car on a scale from one to five. Not surprisingly, I suppose, the "behavioral attractiveness" assessments of the subjects—that is, how they rated the cars—gave the prize to the sports cars by a large margin. Looking at the brain data, the researchers found that when the subjects looked at the sports cars, the ventral striatum and the medial orbitofrontal cortex—brain areas that belong to the reward system—were more active than when the subjects were looking at the small cars. I hope no one is surprised to learn that these reward areas are also activated when male subjects see photographs of female faces. Two conclusions: the most rewarding cars for car buffs are indeed sports cars, and male car buffs look at sports cars with brain responses similar to when they look at attractive women!

Such imaging data reinforce the notion that activity in

the reward system in the brain could be used to measure the appeal of certain products to potential consumers. However, real-life decisions are often more complex than what the activity in reward system areas can tell us. For instance, I may like an expensive sports car, but it still remains expensive. If I can't afford the car, I have to find a way of suppressing the impulse to buy it, or I will be in debt (or too much debt). Generally, these control mechanisms are implemented by cortical areas in the frontal lobe. The desire to buy the car that I know is too expensive creates an internal conflict in me—that is, in my brain. Something has to give. Measuring the activity in these brain areas might tell neuromarketers what it might be.

Other frontal lobe areas also seem to play key roles with regard to our preferences and behavior as consumers. In a famous brain imaging study, Read Montague and his colleagues at the Baylor College of Medicine in Houston used fMRI to look at neural activity while subjects tasted and compared the two most popular sugared drinks. You know which ones they are: Coke and Pepsi. First, the subjects' preferences were recorded with a blind taste test outside the scanner. Then each subject was rolled into the scanner, and for this portion of the experiment, Montague and his team devised a special apparatus that would allow delivery of the cola to the subjects with good control. They certainly couldn't just pop a can of soda and ask the subjects to drink from it! Their solution was clever. They used a computer-controlled syringe pump that injected the colas into cooled plastic tubes held in the subject's mouth with a plastic mouthpiece. The amount of soda was small enough to allow the subjects to swallow comfortably even while lying in the scanner. One final trick was to deliver *decarbonated* drinks, to make sure that the amount of pure cola delivered by the computer-controlled syringe pump was the same at each trial. (I know, one could argue that the fizz effect might somehow be important to the overall taste of the drinks, but let's set that quibble aside.)

The apparatus worked, and the results were clear. When the subjects did not know which brand they were tasting, activity in the medial orbitofrontal cortex correlated with the subjects' decision making. This was the region that seemed to be calling the shots in the blind taste test. And now the test gets really interesting, perhaps because Montague understood the implications of the famous taste tests with Coke and Pepsi many years earlier. In those *blind* taste tests, Pepsi triumphed. When the brand was known, Coke won. This "brand effect" that makes Coke the suddenly better soda is clearly one of the many cultural effects that dominate our preferences for food and drinks. The initial results were replicated in numerous independent tests and, almost inevitably, used by Pepsi in its series of commercials called the Pepsi Challenge. These blind taste tests also always came out in favor of Pepsi.

Montague must have been intrigued by the idea of solving, once and for all, the mystery of the neural origins of brand effects. In a second part of the brain imaging experiment, therefore, his Baylor team delivered the colas into the scanner while also identifying each selection as Coke or Pepsi on a computer monitor. They then compared the activity in the brain during this setup, when *subjects knew the brand*, versus activity in the previous setup, when they did not. The idea was simple: any increased activity in one or more brain areas while subjects sipped the cola while knowing the brand would pinpoint these areas as responsible for the culturally mediated brand effect. And would they be the same areas pinpointed during the blind taste test?

The experiment did pinpoint one area as more active when the subjects knew the brand. It was the dorsolateral prefrontal cortex. This result was not really surprising, because this area is well known for its role in "executive control" over other neural systems. However, it is not the same area as pinpointed during the blind taste test. We conclude, as Montague and his collaborators did, that when brand names enter the picture, the dorsolateral prefrontal cortex overrules the activity in the medial orbitofrontal cortex, which seems to be the evaluation center for taste uncontaminated by the knowledge of the brand name.[4]

The Coke-versus-Pepsi study, together with the study on car buffs, tells us that neuroscience can indeed be effectively used to investigate our evaluation of products and brand names. They also suggest that the activity in brain areas belonging to the reward system, together with regions known to exert executive control over brain activity, seems to be an effective biomarker of subjects' evaluation of products. On the other hand, these studies leave me wanting, because they do not address the fact that most commercials and ads employ actual people, either celebrities or actors, speaking to us and doing things on-screen. I wanted to know what happens when viewers watch these people in action. What happens with their *mirror neurons*? No one had even looked at this question until recently, when we found out the answer.

THE ONE-NIGHTER: "INSTANT SCIENCE" AND THE SUPER BOWL

My hypothesis with regard to mirror neurons and ads is relatively simple. If we look at brain activity of subjects watching commercials, we necessarily find some activity in mirror neuron areas, at least in those commercials in which people are doing things. A *high* activity in mirror neuron areas during such experiments, I propose, represents some form of identification and affiliation. I say this because, as we have seen, one of my hypotheses regarding mirror neurons and social behavior is that the activity in the mirror neuron system is an index of our sense of affiliation with other people. We have seen how these cells in the brain help us understand the actions of others by simulating in our brains those very same actions with the activation of our own motor plans. By doing so, mirror neurons also help us feel what other people feel. Moreover, we saw in chapter 5 that mirror neurons are also concerned with our own process of self-recognition. In short, these cells seem to create some sort of "intimacy" between self and other, and it makes sense to posit that activity in the mirror neuron system may also be relevant to the sense of belonging to or being affiliated with a specific social group whose members, we feel, are more similar to us than other people.

In modern life, affiliation takes many forms. Race comes to mind, of course, and so does nationality. After living in Los Angeles for fifteen years, I still have a strong sense of myself as ethnic Italian. However, there are also more culturally mediated forms of affiliation. I feel that I belong to the worldwide community of neuroscientists, and I may also have a similar (albeit probably a bit weaker) feeling of belonging to social groups defined in a variety of different ways, from Macintosh users to iPod listeners, from tennis fanatics to opera buffs, from wine connoisseurs to sushi lovers. Within some of these spontaneously forming social groups, the feeling of belonging—or at least of sharing an experience—is probably deeper and more meaningful than for others. For instance, it is likely that the feeling of being "a parent of an adolescent" is much more meaningful to most of us in this group than our feeling of belonging to the community of sushi lovers. I think this phenomenon gets heightened even more when it comes to important societal issues: being a liberal as opposed to being a conservative, being pro-choice as opposed to being pro-life. Such identifications are felt deeply by many people. With that last outburst I am getting ahead of myself—we will look at political affiliations and their relationships with mirror neurons in the next chapter. For the time being, I want to focus on identification and affiliation issues relevant to neuromarketing, and again the basic hypothesis is simple: identification with a product as revealed by activity in the mirror neuron system should be a very good predictor of future behavior—that is, future decisions and purchases.

I was mulling over these ideas when Joshua Freedman, a

psychiatrist and co-owner of FKF Applied Research, a consulting firm that explores new tools for advertising research, proposed an experiment to me in the fall of 2005. It involved the Super Bowl. "What is the Super Bowl experience?" Joshua asked rhetorically. It's the football game *and* it's the advertisements between the plays (between *every* play, it sometimes seems). One hundred and forty million fans tune in—that's in the United States alone, with many millions more abroad (though nothing close to the World Cup final). Advertisers spend millions of dollars on special ads, the most expensive and therefore most important of the year. Some of the ads are aired only once. All of them are now an integral part of the whole show. A lot of those 140 million viewers at Super Bowl parties are as interested in the ads as in the game. (Quick, who won in '06? Or take your time, it probably won't matter. Many viewers can't answer the question a month after the game, much less a year later.) The ads are one of the most talked-about topics in the country the day after the game. Marketing experts and their focus groups rank the most effective, the most entertaining, the most surprising—and the worst, of course. The ads are heavily played on YouTube the next day. Viewers can cast a ballot on any one of several websites.

The Super Bowl ads are *big*, and Joshua proposed an imaging experiment performed practically during the game. Our subjects would watch the ads while we watched their brains. In order to analyze the data rapidly, we would need all the computing power available, because brain imaging datasets are quite large and their analyses require intense computations. If we could pull this off, Joshua gloated, we would have a different kind of ranking of the ads, one based on quantifiable brain activity, not just expressed *opinions*. I found the idea amusing, but I reminded my friend that that's a big "if." We generally require several months, and not infrequently *years*, to complete a complex imaging experiment, from the first glimmer of the idea to the last analysis of data, and Joshua wanted to get everything organized within a few months. Then—the biggest problem by far—he wanted to complete the imaging data acquisition and the computations to process the data overnight. In the end, however, I agreed. When I subsequently wrote a report about our experiment for *Edge—The Third Culture*, the online salon of John Brockman (www.edge.org), I dubbed this "instant science"—of which there are very few instances.

How on earth could we manage to pull this off? Well, for starters, we would use a limited number of subjects—only five. As a brain imager, I would have loved many more, but time would be truly of the essence. (We repeated the Super Bowl experiment in 2007 and managed to double that number.) We would need volunteer subjects in their late twenties and early thirties, because that is the demographic often targeted by the Super Bowl ads, and we would need to insulate them from the live broadcast. They had to see the ads for the first time when they were in the brain scanner. And most important of all, we would need to obtain a copy of the ads practically instantaneously. They were not available before the game, and we could not wait until after the game to select the ones we wanted to use. That would have been too late. In the end, we decided to tape and digitize the ads shown during the first half of the game, and start running our subjects during the second half. This decision brought up another complication: not all of the ads on the broadcast are new "Super Bowl ads," and we would have no idea when those would be aired. Joshua solved this issue by having a large group of FKF Applied Research personnel follow the game, monitor the ads as they aired, and confer online. The tape was rolling, so to speak, during every ad segment, and the genuine Super Bowl ads, along with some ordinary "control ads," were immediately digitized and made ready for feeding through our computers in the brain imaging lab to the scanner, where our subjects would view them wearing the standard high-definition goggles. (The idea behind the ordinary "control" ads was to make certain there was nothing "magic" about the Super Bowl ads, that they do not get the reward areas in the brain fired up while ordinary control ads do not. We did not actually think there would be, but we also thought it was a good idea to have some data to confirm or undermine our intuition.)

Another problem: the files that contained the digitized ads would be too large to transfer through the Internet, so Joshua would arrive at the Brain Mapping Center shortly after the end of the first half with the hard disk, ready to load the files into our computer. We had never done experiments on the fly like this. If anything, the practice of science is exactly the opposite of doing experiments on the fly. There were about eight of us working on the experiment in one capacity or another, compared with a normal contingent of two or three. Except for the subjects, we all were stressed-out. (The subjects had nothing special to do while waiting their turn, or when they were inside the scanner, where all they were asked to do was watch the ads "broadcast" in their goggles.) Lots of little things went wrong throughout the afternoon and evening and delayed us, and the presence of a crew from ABC's *Good Morning America* and a reporter from the *Los Angeles Times* did not help. They were following our every move and noting our every problem. The first half of the game ended around 5:00 p.m. on the West Coast. We managed to complete the brain scan of our first subject around 6:30 p.m. Our plan was to start analyzing the data of the first subject almost instantly, while rushing the second subject into the scanner. With these parallel procedures, we were hoping to finish the data analysis late that night. And we did, a few minutes before midnight. We were completely exhausted. If I had ever had any doubts, that night erased them completely: it is much better, at least for the scientists, to do experiments at their own pace rather than at the pace dictated by external considerations. Instant science is not the way to go. But I also had high hopes that all the trouble this time would be worth it.

MIRRORING ADS

We first established a baseline with the subjects in the scanner, measuring their brain activity while they "did nothing" (that is, watched a blank screen or kept their eyes on a fixation cross in the center of the monitor). This resting baseline allows us to compare such resting activity with the brain activity measured while subjects perform the assigned tasks. (This baseline was not relevant in the previously discussed experiments.) When we projected the ads themselves into the subjects' goggles, we measured activity in the whole brain, with special attention given to four key neural systems: the mirror neuron system, the reward system, the brain centers for executive control, and the emotional brain centers. In the analysis that followed, we first looked at the overall activity in these systems during viewing of all the ads, as compared with the resting baseline. We also looked at sensory brain areas for vision and sounds, because watching the ads required *watching* the ads and *listening* to their sounds, words, and music. We did not really consider these areas theoretically interesting in this experiment, but we decided it would be useful to compare their activity with the activity in our neural systems of interest. Indeed, that's what we found: nothing special in the brain areas for vision and sound. They were consistently activated by each ad in each subject. This was not surprising, but it was reassuring. After all, we had never done such a wild experiment, and it was good to see in our data the features we would typically observe in more mainstream experiments.

When we turned to the activity in the four neural systems of interest, we found that for several of the ads, the reward system, the executive control system, and the emotional brain centers did *not* show any change compared with the baseline, nor did they demonstrate reduced activity. This held for every subject, and it was somewhat surprising given that the ads were presenting highly desirable objects or services in supposedly engaging ways. That's certainly what the sponsors were paying top dollar for. Only one system was consistently activated compared with the resting baseline in each subject and for each ad. This was the mirror neuron system. It never failed to activate. Clearly, the presence of people (actors) in the ads was the main reason. Intriguing to me, however, was the variation in activation of the mirror neuron system. The activity for some ads did *not* seem related to obvious physical aspects of that specific ad (hand gestures, for instance, or lots of people performing various actions). My hypothesis is that the higher mirror neuron activity observed in certain of the ads was due to higher levels of identification for those ads from the viewers. The Super Bowl data cannot really say much about this hypothesis, but other imaging data collected in my lab do seem to support it. In a completely independent experiment, we imaged brain activity of people who are and are not cardholders of a certain credit card. All subjects were shown various pictures of people busy with their shopping. Some of them, including cardholders and non-cardholders, saw these pictures with the logo of the credit card superimposed in the lower-right corner. Others saw the pictures without this logo. The results were striking: Among the *non*-cardholders, mirror neuron activity did *not* reflect the presence of the logo at all. These *non*-cardholders didn't care about the logo one way or the other; they didn't identify with it. Among the cardholders, however, there was higher activity in mirror neuron areas while subjects were watching the pictures with the logo of their credit card, compared with watching the pictures without it. Could it be that the activity in mirror neuron areas simply reflected the simulation of holding the card, as if the subjects were thinking, "I've held my card the same way"? Not really, because the pictures we used did not show people holding the card. In these subjects, we had probably imaged the neural correlate of a phenomenon of "identification" that is mediated by mirror neurons. It is as if these cardholding subjects, while watching the pictures with "their" logo, were thinking, "Those people are like me."

The other striking feature of the data from the Super Bowl experiment was the dissociation between the behavioral data (each subject was interviewed immediately after emerging from the scanner) and the brain data. When explicitly asked to name their favorite and least favorite ads, the subjects came up with very specific choices, but their brains were telling us a different story. The ads named as the best often gave almost no response, or a weak one, in terms of those neural systems we considered more meaningful for human behavior. Other ads that gave robust responses in the mirror neuron system and in the reward system were not mentioned by the subjects. Although one could interpret such dissociation between the brain data and the verbal reports as suggesting that the brain responses may not be reliable indicators of the future choices of the subjects, I tend to favor a different interpretation. As we have seen in the experiment with the pretty faces on the cards and in the case of translational dissociation discussed earlier, people tend to be quite out of touch with their own choices, and their verbal reports on how they made decisions are unreliable. Brain responses in key neural systems that research has indicated as highly relevant to hu-

man behaviors such as motivation (reward system) and empathy and identification (mirror neuron system) are in my
opinion better predictors of future subjects' choices.

Obviously, I cannot prove my hypothesis. The only way
my hypothesis could be seriously tested is with an experiment
that, even though eminently doable, is unlikely to actually be
done; it would require a serious and committed partnership
between neuroscientists and the business world, which is unlikely to happen in the very near future. But for the fun of it,
here's how I would test my hypothesis that brain markers are
much more reliable indicators of consumers' future purchases
than are their verbal reports. First we would need to identify
two states in the United States that are relatively insulated
from each other and yet have at least one sector of the population that is comparable with regard to the most important
demographic factors. Next, we would identify *one* target product that is marketable to this one sector of the population in
both states, and then prepare a series of radically different
advertisements for the product. Finally, we would run focus
groups *and* brain imaging experiments on a large group of subjects in each state to analyze reactions to each ad. As we did
with the Super Bowl ads, we would subsequently identify the
ads showing a considerable dissociation between the verbal
reports and the brain imaging data—positive verbal reports
but not much happening in the reward and mirror neuron systems, and negative verbal reports but powerful brain activity.
In one state we would now run *just* the ads that received good
verbal ratings but "bad" brain responses, and in the other state
we would run *just* the ads that received bad verbal responses

but "good" brain responses. If my hypothesis is correct—that is, if brain indicators are more reliable indicators of future choices than what people tell interviewers—the advertised product should see more sales in the state whose residents were inundated with the ads that received the "good" brain responses.

Easy, no? I'm completely serious about the setup, and I don't think it would be a difficult experiment to perform once the shared demographic in the two otherwise isolated states was carefully identified. The only other real hurdle I see is finding a large corporation whose leadership understands the power of what's going on in neuroscience today and is therefore willing to make the investment, which for such a corporation would be a pittance anyway. I know some neuroscientists who would love to do their part. There are many others.

Another question this thought experiment would answer is whether ads actually work in the first place. In spite of the large amount of money spent on advertising, I believe that nobody can convincingly claim to know the answer—with the exception of one field in which people are convinced that ads work, or at least one special type of ad. I am talking about political consulting, where the pros believe that negative ads work and have at hand some classic examples to support this belief. One is the famous ad for Lyndon Johnson in the 1964 presidential campaign against Barry Goldwater, showing the little girl picking the petals from the pretty daisy while a mushroom cloud explodes in the background. Unstated message: Goldwater is a trigger-happy warmonger. Result: Landslide Lyndon. Another is the equally famous "revolving door" ad for George H. W. Bush in his 1988 presidential campaign against Democrat Michael Dukakis, governor of Massachusetts. At one point that summer, the polls gave Dukakis a lead of 17 points. Then the Bush campaign ran the ad featuring Willie Horton, a black man who had been in and out of prison in Massachusetts. The ad played explicitly on racial fears and crime fears. Result: the first George Bush wins easily. There are many more examples that contribute to the received wisdom in politics that "going negative" wins, that no matter what people tell pollsters and exit pollers about the reasons for their votes, they in fact vote their anger and their fear.

Is there any brain data that may support this idea? I believe we found it in my lab, even though we were not really looking for it. Most important, I believe that this brain finding should make people think twice about the long-term value of negative ads.

THE EFFECTS OF NEGATIVE ADS

In the spring of 2004 we started a new experiment measuring brain activity in the "partisan brain" while registered Democrats and Republicans watched a series of pictures of the three presidential candidates that year, incumbent Republican President George W. Bush, Democrat John Kerry, and the independent Ralph Nader. We scanned approximately half of the twenty subjects in the spring, after Kerry's victories in the primaries made his nomination a foregone conclusion. Getting wind of this study, the *New York Times* reporter John Tierney interviewed me. His front-page article on the experiment, published on April 19, was titled "Politics in the Brain?" Suddenly our imaging experiment became probably the most media-covered brain study ever. Major TV networks picked up the story. This was all astonishing because we had not even finished the experiment. Shades of "instant science"!

Media coverage for a scientist is generally a good thing. It means that the experiments we do are interesting to the population at large, and that an understanding of what we do does not require a Ph.D. in neuroscience. In the lab at UCLA, we were indeed pleased by the attention in this case. However, we did not realize that a high-profile story can be a double-edged sword. (Naïve, I realize.) Indeed, the coverage made it basically impossible to recruit new subjects who did not know about the experiment and, most important, did not know about the results of our preliminary analyses, which were reported in Tierney's article in *The New York Times*. This was a major problem relating to what we call metacognitive processes. Simply put: the subjects' prior knowledge would screw things up for the experiment. What could we do? Wait. Stop recruiting subjects for a while. People do not have long-term memory for these kinds of things, and the media would be feeding them a never-ending series of new stories and new topics to debate. There is no shortage. The supply seems infinite. We thought that by stopping recruiting for about three months, we could effectively solve our metacognition problem. Indeed, when we resumed our experiment in the summer, people had basically forgotten what had been reported.

At best, they had vague memories about some kind of brain experiment at UCLA.

During the summer, with the election approaching, media interest picked up again. But I had learned my lesson. This time I stipulated an agreement with all media outlets who wanted to interview me. They would not run the story until we had completed our data collection, which we did in early September. At this point the experiment as originally planned was completed, with the peculiarity that approximately half of the subjects had been studied in the early spring and the remaining half in the late summer. In this study, our subjects were not only registered Democrats or Republicans; they had also made up their minds about the election, not surprisingly, since this was one of the most polarized presidential elections in American history. Such partisans are able to look at the very same data and reach completely opposite conclusions. This we all know. But *how* do they do it? This we do not know (or did not, during the experiment). We hoped to help answer the question by using very simple stimuli—the faces of the three candidates—with the hope of minimizing the complex factors that may be at play during a political debate and reaching the *core* of the feeling of affiliation with a party and a candidate. My main hypothesis was that feelings of empathy and identification for your own candidate may be supported by activity in mirror neuron areas, or by super mirror neurons.

In the subjects studied in the spring, the brain activity we observed was right in line with our main hypothesis. While watching their own candidate, both Democrats and Republicans activated the medial orbitofrontal cortex. As we know, this brain area is typically activated by rewarding stimuli—for instance, one's favorite food. Our candidates were people, not inanimate food, but this brain area is also associated with positive emotions such as happiness. Watching one's own candidate is obviously associated with positive feelings. Most important, our depth electrode recordings in the epileptic patients described in chapter 7 suggest that in this brain area there are super mirror neurons. Therefore the neural activity observed in this region was consistent with the hypothesis that watching one's own candidate would elicit feelings of empathy and identification through the mirror neuron system.

So far, so good. To our vast surprise, however, the subjects we studied in the late summer did *not* show the medial orbitofrontal activity while watching their own candidates. How was this possible? We looked at the data very carefully, but the numbers refused to change. The only reliable difference between the brain responses of the early-spring and the late-summer subjects was indeed in this one region, the medial orbitofrontal cortex. It was activated in the spring, dormant in the late summer. This deactivation suggested not that it simply failed to activate, but rather that an active process of shutting down this region was going on in the late-summer subjects while they were watching the pictures of the candidates they planned to vote for.

I believe this dramatic change in the medial orbitofrontal cortex in the late-summer subjects was due to the change in the political climate and the heavy use of negative ads and personal attacks with which both sides had pummeled the opponent throughout the summer. In such a toxic climate, how could you possibly identify and empathize with your own candidate, even though he would still receive your vote? It was almost impossible. The campaign had tainted all of the candidates, even for their partisan supporters.

Obviously, this interpretation is entirely post hoc. However, if we consider what was going on in that campaign and what we know about the medial orbitofrontal cortex, my explanation is reasonable. And if it is correct, I believe this is bad news for our society. It obviously shows that negative ads work (no news to political consultants), and it also shows that negative ads can create a dangerous emotional disconnect between voters and the leaders who should represent them. A healthy democracy, in my opinion, needs mechanisms of empathy and identification between the people and their political representatives. Without these unifying emotions, we run the risk of an ever-growing disenchantment with the political system that may make people more receptive to other forms of government. And thus far, the alternatives to the modern democracies have proved to be much worse than what we have now.

TEN

Neuropolitics

THEORIES OF POLITICAL ATTITUDES

In the late 1990s, Darren Schreiber, then a UCLA graduate student in political science, now a political science professor at the University of California, San Diego, approached the faculty of our Brain Mapping Center with the idea of testing certain theories about how political attitudes are formed. At that time, the use of brain imaging for such a purpose was basically unheard of. Now, if not quite mainstream, it is not that unusual. A few labs are doing such research, ours included. Inevitably, of course, as Darren put his idea into action and other experiments followed (including the 2004 election experiment I just described), I began to wonder whether mirroring, and therefore mirror neurons, play a role in all this. Typically, serious students of politics have liked to believe that political thinking is a highly rational process in which automatic mirroring should not play a major role. However, we have seen how mirroring is a pervasive form of communication and social interaction among humans. Given that a major component of politics is affiliation with others with whom we share values and ideas about how society should be organized, I think forms of mirroring are almost certainly involved in some aspects of political thinking.

And exactly how rational is political thinking to begin with? That's what Darren wanted to find out, because data from national surveys had stirred a long-standing debate in the political science literature. When citizens were asked a variety of questions on political issues, a clear pattern emerged. With those subjects who responded quickly, the responses were consistent in terms of political attitudes. They "made sense." For instance, if one of these quickly replying subjects expressed a "liberal" attitude on abortion, the same subject would probably respond with a "liberal" attitude on education or gay rights. However, another group of subjects required, on average, quite a long time to respond to the questions, and their answers were not consistent. On some questions they would have the "liberal" attitude, on others the "conservative" one. Nor was there any consistency within the group: the same question would elicit a liberal answer from some of the slow repliers, a conservative answer from others.

Overall, the results from these surveys seemed to identify two different kinds of citizens. Was there any major variable that could easily differentiate between them? The answer seemed to be yes. The subjects who knew a lot about politics were the ones who responded quickly and with consistent attitudes. The subjects who didn't know nearly as much took a long time to respond and then did so "inconsistently." In the 1960s, the political scientist Philip Converse wrapped up his analysis of this phenomenon by suggesting that political sophisticates had well-informed although rather crystallized political opinions, whereas political novices had no opinions at all, and when responding to political survey questions, they basically flipped coins. Perhaps this summary sounds rather mundane today, but it started quite a controversy in the political science literature. About ten years after Converse's proposal, another political scientist, Chris Achen, argued that the political novices simply were not able to map their true attitudes during these political surveys. Their seemingly inconsistent responses were due not to a lack of political attitudes, but to imperfect, inadequate political surveys. And a third hypothesis was more recently proposed by John Zaller (incidentally the mentor of Darren Schreiber during his graduate studies) and Stanley Feldman. They suggested that the novices' inconsistent responses were not due to a complete lack of a political attitude—or to the artifact of imperfect surveys. They proposed instead that while the crystallized opinions of political sophisticates were based on an almost automatic retrieval of facts and prior considerations, political novices retrieved information relevant to the political questions as they went along with the survey. Only the more salient information—generally speaking, the latest news—determined their answers. These novices *do* need a weatherman to know which way the wind blows! This is why they seemed to flip coins when answering the survey questions.[1]

If Zaller and Feldman's hypothesis is correct, the difference between political sophisticates and political novices is mostly due to cognitive differences stemming from different levels of expertise, the same differences that would be noted between so-called sophisticates and novices in any field. Sophisticates are engaged in a well-practiced task, the novices in a new one. In fact, brain imaging data showing strikingly different patterns of activation between a well-practiced task and a novel task have been around for years.[2]

Darren Schreiber set out to use brain imaging to look at all of these questions about political thinking. I thought he had a very clever idea, but I have to admit that my interest was also motivated by my research on mirroring. With politics, the sophisticates are almost junkies. They're hooked, thanks in large part to the endless opportunities provided by the Web. I wanted to find out whether a political junkie's brain would produce higher mirroring responses while watching politicians compared with watching other famous people. I believed that it would. Darren, his mentor John Zaller, and I met several times over the course of a year to figure out how to set up a series of experiments that would address the various issues we were interested in. We were venturing into the unknown. There had never been a brain imaging experiment on issues of political science. It took us a while to shape our interests and ideas into viable experimental designs. When we finally did, we had to face the standing problem in neuroscience (and almost every other field). How could we fund the project? Imaging is an expensive scientific enterprise. Use of the MRI alone, without taking into account overhead, salaries, volunteer fees, and so on, is typically about six hundred dollars per hour. Total cost for our imaging experiments varies from tens of thousands to hundreds of thousands of dollars. We designed Darren's study to keep costs at a minimum, but it was still significant money, and interdisciplinary projects like ours are almost always the most difficult to finance, because it is difficult to identify funding agencies who will also be interested in breaking down barriers. Most spend their money within very tight parameters. Luckily, at UCLA we have a research funding opportunity called the Chancellor's Fund for Academic Border Crossing, specifically designed for interdisciplinary projects involving two professors from different disciplines mentoring a graduate student who wants to perform interdisciplinary work. In the summer of 2000 we applied for this funding. Coincidentally or otherwise, we received the good news on election day that fall. We thought this was a good sign. Then the electoral mess in Florida dragged on and on, and we could only hope our experiment would proceed more expeditiously.

MIRRORING AND THE POLITICAL JUNKIE BRAIN

To maximize Darren's chances of obtaining an experimental effect, we thought it would be useful to select subjects at the two ends of the spectrum. Among the sophisticates, we wanted those most knowledgeable on subjects in politics. Among the novices, we wanted to recruit the most clueless individuals, who knew nothing and were content not to. Darren got down to business and began recruiting subjects in the early months of 2001. To select these individuals, he had prepared an extremely detailed series of questions. His screening interview would take some hours for each subject. To find the ideal sophisticates, he interviewed stalwart members of the Democratic and Republican clubs on campus, and he quickly found the "political junkies" we were looking for. These young men and women were well-informed, and their political attitudes were radical and crystallized. Darren's sophisticates looked like ideologues.

The recruitment of the novices was not terribly painful either. Darren advertised the study through the usual recruitment channels, and I don't suppose anyone will be surprised to learn that lots of UCLA students did not (and still do not) know much about politics. Darren had plenty of novices to choose from. The subjects he chose were indeed clueless and utterly without well-formed political attitudes. They knew that Bush was the new president, they knew there had been some kind of problem on election day, they might even have responded to the phrase "hanging chad," but that was about it. (Today, they would also know that Schwarzenegger is the governor of California.)

A secondary goal of the interviews with the novices was to gather the information necessary to design one of the imaging experiments. One key for Darren's design was that the novices had to at least recognize the faces of the politicians, even if they knew almost nothing about them, so he explicitly asked his potential subjects whether they recognized certain faces. This is how we discovered the depth of the cluelessness on the UCLA campus. The face of Joe Lieberman, who had been Al Gore's running mate in the famous disputed election less than a year earlier, was basically unknown among the mass of students. We also factored in another variable, the constant hot-button issue in American politics: race. The whole experimental design thus comprised three different kinds of faces: political or not political, famous or not famous, and white or African American.

On scan day, the subjects were simply asked to watch the faces while we were measuring their brain activity with fMRI. We found what I had predicted with my theory that mirroring indicates, among other things, a sense of affiliation, of belonging to a specific community within the larger community of our society. Politically sophisticated subjects had higher activity in mirror neuron areas when they viewed the famous political faces, compared with when they viewed famous nonpolitical faces and unknown faces. The political novices did not show any such difference in mirror neuron areas when they were watching political and nonpolitical faces. When we compared the results obtained with the political sophisticates with the results obtained in our previous study on imitating and observing facial emotional expressions described in chapter 4,[3] we found remarkably similar locations of activation. This anatomical correspondence suggests that even for the more abstract types of mirroring I had hypothesized as the basis of these activations—the sense of belonging to a specific community—the mirror neuron system still uses the basic neural mechanism that also activates during more mundane mirroring tasks.[4]

The experiment using the photographs to look for mirror neuron activity among political sophisticates was one of two Darren conducted with the same subjects. In the other one, he tested whether sophisticates and novices use different brain areas when thinking about political issues. His "expertise" hypothesis had suggested that they do, because previous data from this kind of imaging experiment—looking at novel versus well-practiced tasks—had shown brain activation in largely separated brain areas. The activations for the novel tasks suggested that they are performed (because they have to be) with a high level of cognitive effort, specifically with enhanced activation in the dorsolateral prefrontal cortex, an area known for its role in the so-called executive functions. On the other hand, well-practiced tasks seem to be performed mostly using information retrieved from memory, using areas in the temporal lobe, an important brain structure for memory. According to Darren's hypothesis, therefore, political novices and sophisticates should show analogous patterns of activation: cognitive areas for the political novices, for whom thinking about politics would be cognitive work, and memory areas for the sophisticates, who already have their answers to political statements and have merely to retrieve them.

In this setup, the subjects listened to a series of digitally recorded statements, half of them political, half nonpolitical. The political statements concerned typical hot-button issues in American politics, and the subjects were asked to agree or disagree with each statement. The statements were carefully crafted so that the initial phrase was always the same. For instance, the political statements started, "The government in

Washington . . ." The final part of the sentence presented the novel opinion for each statement—for example, "should encourage adoption by banning abortion." These loaded statements were relatively similar in structure to the questions Darren had used to reveal the different patterns of behavior between political sophisticates and political novices. The specific and careful form of presentation allowed us to deliver the critical part of the statement in a relatively well-defined temporal window, which helped us look in a fairly precise way at the brain changes occurring from the presentation of the important material to the response of the subject, which was given by pressing one of two buttons.

Darren got his answer loud and clear. The pattern of brain activation for political sophisticates and novices was quite different—but not as we had expected. To everyone's surprise, the results did *not* show the expected cognitive/memory distinction. The two areas that demonstrated the striking dissociation between sophisticates and novices were the precuneus and the dorsomedial prefrontal cortex. Both belong to a neural system called the default state network, which had been discovered only recently by Marcus Raichle and his colleagues at Washington University in St. Louis.[5] The default state network is a peculiar set of cortical areas that have high activity while the subject is resting and doing basically nothing, and reduced activity while the subject performs cognitive tasks. This reduction in activity was substantially independent of the kind of cognitive tasks the subjects were performing. All in all, this was a bizarre neural response that was difficult to interpret. By looking at certain physiological parameters measured with PET (the now somewhat out-of-favor technique that uses radioactive material, as discussed), Raichle and colleagues demonstrated that these regions were actually *shutting themselves down* during a variety of cognitive tasks. Thinking carefully about this, they suggested that these areas represented some kind of default state of the brain that is dominant when there are no specific goals or tasks at hand, when the subjects (that is, we humans) daydream or "do nothing." When certain tasks require attention, this "default state" is overridden and its network shuts down.

This analysis ties in perfectly with the results from Darren's test. During the political questions, these "default state" brain areas were activated in the sophisticates, who think about politics all the time (their own "default state") and do not need to deploy attention to the political statements. They need only their memory banks. The novices, however, had to think about the political statements, so they geared up for cognition and shut down the default network.[6]

To judge by the brain imaging literature, increased activity in these default state areas is extremely rare during any kind of task. As it happens, we had previously observed one of the most robust, if not the most robust, increases in my lab.[7] Very provocatively, this increased activity in the default state network was paralleled by increased activity in mirror neuron areas. And now Darren's experiment had picked up increased activity in the default areas for the political sophisticates. Is there a link between the results of these two experiments? More generally, what is the relationship between mirror neuron areas and default state areas? Before we consider these questions, let's look at that prior study, which was unique, even apart from its results, partly because the driving force behind it was an anthropologist, not exactly the kind of scholar who typically participates in a brain imaging study.

BRAIN POLITICS

Alan Fiske is an anthropology professor at UCLA who has performed a detailed ethnographic analysis of the Moose people of Burkina Faso, a West African society. Drawing from this fieldwork and from scholarly work encompassing a variety of disciplines studying a variety of cultures, Alan proposed a model of human social relations, according to which we relate to each other using four elementary forms of social relations: communal sharing, in which people have a sense of common identity; authority ranking, in which people relate to each other following a hierarchy; equality matching, in which there is an egalitarian relationship among peers; and market pricing, in which the relationship is mediated by values that follow a market system. Alan contends that these four elementary relational structures and their variations account for all the social relations among all humans in all cultures.[8]

Alan published that work in 1991. Eight years ago (about a year before Darren Schreiber walked into my office), Alan contacted me about teaming up on an imaging experiment relating to his well-known model of social relations. I found the idea fascinating because he made me realize that those of us in the lab were basically studying responses in mirror neuron areas while subjects were simply watching or imitating individual actions. These actions were rarely surrounded by a social context. In the few instances in which we used a social context surrounding the action, as in the "intention" experiment with the teacups described in chapter 2, the context was composed only of objects, no people. Given our claim that mirror neurons were important neural elements for social behavior, I knew it was important to measure brain responses in mirror neuron areas in an experiment in which the observed actions were highly relevant to human social relations. Talking with Alan about his idea, I envisioned an experimental design for a study that could suit both of our purposes: the only task for the subjects in the scanner would be the observation of social relations between people. Of course, we could not bring a bunch of people into the scanner room and stage various interactions while our subjects watched, so we prepared a set of video clips depicting everyday social interactions. In order to simplify the experimental design, we also decided to focus on only two of the four relational models of Alan's theory. Once again, we were moving into uncharted territory. In these cases, a relatively simple experimental design is highly advisable.

As in the brain imaging experiment on politics performed by Darren Schreiber, in which we picked subjects at the far ends of the political continuum, we picked the two social relational models that seemed at the far ends of a continuum. One was communal sharing, predominantly based on kindness and sharing, and the other one was authority ranking, based on hierarchical inequality. The tricky issue was that communal sharing relations seem inherently positive, while authority ranking relations are typically perceived in a negative way, especially by North American subjects. This was a "confounding factor" that we had to control for if we were to achieve a pattern of brain activation that truly reflected differences in the way we process social relations, not differences in how Americans feel about authority figures! We ended up with thirty-six video clips, a fairly large set for such an experiment, half depicting communal sharing social relations, the other half depicting authority ranking social relations. Some of the clips for each relationship clearly elicited positive emotions, the others elicited negative emotions, thus controlling for the "emotional valence" of the clips.

We used screenwriters, actors, and one director to actually make the video clips. Finding these professionals was not much of a problem, given that we live in the entertainment capital of the world. One of the many fascinating aspects of this experiment was explaining to the screenwriters the anthropological model that inspired it and working with them to create realistic scripts that would depict the social relations in a variety of everyday situations. The "development" process, as they say in that trade, was fairly involved, but after about six months we were satisfied with the scripts. We brought in the director and the performers and shot a series of very short movies. Each story was identically structured, introducing one character for "baseline" purposes, then bringing in the second character for the interaction—the "relational" segment. The depicted situations were widely variable, from

office scenes to basketball courts, from lovers playfully interacting to judges ruling in court.

Looking at the brain data of the subjects watching these
scenes, we found robust activity in mirror neurons, as expected, because the observed characters were making all sorts
of actions during the course of the scene. Indeed, mirror neuron activity in this study seemed stronger than anything we
had previously measured, and this robustness was especially
high during the relational segment of the clip. This correlation confirmed that mirror neurons are especially interested in
actions that unfold during social relations, probably because
those actions are critical to our understanding of the relationship. Other brain areas also demonstrated fairly robust activity while subjects watched the social interaction clips:
particularly the default state network, which had been impli-
cated in Darren's experiment with political junkies answering
political questions. My interpretation of these data is that
while political junkies think about politics all the time (it's
their "default state"), most people think about social relations
all the time (it's our "default state"). Who am I? I am the husband of my wife, the father of my daughter, the son of my parents, the mentor of my trainees, the colleague of my peers,
and so on. I am constantly defined in relation to other people.
It seems that there is, in addition to the mirror neuron system,
another neural system in the brain—the default state network—that is concerned with both self and other, in which
self and other are interdependent.[9] While mirror neurons
deal with the physical aspects of self and others, I believe the

default state network deals with more abstract aspects of the relationship between self and other—their roles in the society/community they belong to.

I am convinced that understanding the fundamental connections between self and other is essential for understanding ourselves—my "two sides of the same coin" argument. Mirror neurons are the brain cells that fill the gap between self and other by enabling some sort of simulation or inner imitation of the actions of others. We are left with one final question: Why on earth do we need to simulate in the first place?

ELEVEN

Existential Neuroscience
and Society

MIRROR CELLS BETWEEN US

For the most part in this book, I have been describing the details of the empirical research on mirror neurons and the implications that flow from that research. We have seen that mirror neurons in the monkey brain are concerned with the control of certain fundamental actions in the animal's motor repertoire, such as grasping objects, biting food, and making communicative facial expressions. They also have the surprising property of firing when the monkey is not moving at all and is simply watching somebody else making those actions. Mirror neurons also respond to sounds associated with actions such as breaking a peanut, even when the action is not seen. Mirror cells fire even when the action is partially occluded, and they are able to differentiate between two identical grasping actions made with different intentions. Taken together, these cells seem to "mimic" in the observing monkey's brain the actions and the intentions of other individuals.

Building on and paralleling the research on monkeys, brain imaging and magnetic stimulation data on humans have revealed a mirror neuron system that fulfills the same functions as it does in monkeys. In humans, however, its role in imitation is even more critical because imitation is so fundamental for our exponentially greater capacity for learning and for the transmission of culture. Human mirror neuron areas also seem important for empathy, self-awareness, and language. We have been working with mirror neurons for barely fifteen years, but already we have learned that these cells are quite likely to be vitally important for our overall understanding of the human brain and mind and, therefore, of ourselves.

All of these repercussions flow from the "simple" mechanism by which mirror neurons fire not only during our own actions but also during the observation of the same actions by others. The mirror neuron system seems to project internally (psychoanalysts would say "introject") those other people into our own brains. How astonishing should these findings seem to us? One of my collaborators—the neurosurgeon Itzhak Fried, who did the groundbreaking work with depth electrodes implanted in the brains of epileptic patients requiring surgery—told me a story at a meeting we both attended. Itzhak operates on patients both in Los Angeles and in Israel. A couple of summers ago, while he was in Israel, he saw on television the presentation of an award to a famous Israeli actor. In his acceptance speech, this actor mentioned mirror neurons. The story, as Itzhak related it to me, is that the actor said that neuroscientists had discovered these brain cells that fire when one makes an action or a facial expression, and also when one observes somebody else making the same action or the same facial expression—describing the basic facts, in other words. He then noted that while neuroscientists found this property extraordinary, they should have asked "us actors," who have known—or, better, "felt"—all along that they must have something like these cells in their brains! When I see someone with a painful facial expression, said the actor, I feel her pain inside me.[1]

When we think about it, the Israeli actor is clearly right. In many cases, this truth seems almost self-evident. When we introspectively look into ourselves, we find this immediate perception of the actions and emotions of others. So why did (and do) neuroscientists find mirror neurons such extraordinary cells? I believe it goes back to the assumptions discussed at the beginning of the book, assumptions we all tend to make that dictate the way we view the phenomena we observe. The most dominant view in thinking about the mind—at least in Western culture—originates from a position that goes back to the French philosopher Descartes and looks at the starting point of the mind and the self as the solitary, private, individual act of thinking, the famous *cogito* of *cogito ergo sum*. Some philosophers have argued that all sorts of problems arise if one accepts these premises, including the famous problem of other minds, which has come up in several contexts here. However, some other philosophers, among them Wittgenstein, certain existential phenomenologists, and certain Japanese philosophers, have challenged the idea that the problem of other minds is a difficult one by emphasizing the immediacy of our perception of the mental states of other people. Remember Merleau-Ponty's dictum "I live in the facial expression of the other, as I feel him living in mine." And now listen to Wittgenstein: "We *see* emotion . . . We do not see facial contortions and *make the inference* that he is feeling joy, grief, boredom. We describe a face immediately as sad, radiant, bored, even when we are unable to give any other description of the features."[2] Mirror neurons seem to explain why and how Wittgenstein and the existential phenomenologists were correct all along.

In this final chapter, I turn to these more straightforward theoretical implications of the discovery of mirror neurons, of which I believe two are the most important. The first concerns intersubjectivity, which has already generated quite a large literature. The second implication has been discussed much less, but I believe it may be even more important. It has to do with the role of neuroscience in shaping and changing our society for the better.

THE PROBLEM OF INTERSUBJECTIVITY

Intersubjectivity, the sharing of meaning between people, has always been perceived as a problem in classical cognitivism. Simply put (very simply—long books have been devoted to the subject): If I have access only to my own mind, which is a private entity that only I can access directly, how can I possibly understand the minds of other people? How can I possibly share the world with others, and how can they possibly share their own mental states with me?

One classic solution to this problem has been provided by an argument from analogy, which goes as follows: If I analyze my own mind and its activity in relation to my own body and its actions, I realize that there are some links between my mind and my body. If I am nervous, I may sweat, even though it is not hot. If I am in pain, I may scream. So far, so good, and I now take this understanding and look at the other person and find an analogy between *that* body and my own body. And if there is such an analogy, there may therefore be an analogy between the other person's body and the other person's mind. So if I see the other person sweating when it is not hot, I may conclude that the other person is nervous. If I see the other person screaming, I may conclude that the other person is in pain. By analogy I arrive at the conclusion that his behavior must somehow be the clue to understanding his emotions and what is going on in his mind.

Although this kind of analogy does not allow me to be completely positive about the mental states of other people, and it does not allow me to share their feelings and experiences, it certainly does allow me to conclude with reasonable certainty that people have minds like my own.

Or so it would seem. The argument has been heavily criticized by some thinkers on the grounds that this way of reasoning about the mental states of other people is way too complex for something we seem to accomplish so naturally, effortlessly, and quickly all the time. Indeed, it reminds us of the inferential approach to the understanding of the mental states of other people proposed by theory theory, as we have seen in chapter 2.

A different kind of criticism of the analogy argument that is less commonly raised—but which I find quite compelling—concerns the overestimation of self-knowledge implied by the argument. As we saw in chapter 9, we are much less in touch with our own mental processes than we would like to think. Remember the phenomenon of translational dissociation discussed in chapter 9, or the experiment on choice blindness, where subjects literally made up reasons why they had chosen a face on a card as more attractive than an alternative, even though it was the alternative that they had actually chosen! How can we use our understanding of the self as a model for understanding other people if we have such limited self-knowledge? Logically, we can't—yet clearly we do, since we successfully predict and explain the behavior of others count-less times each day. We must do this through some process other than making inferences based on an abstract analogy between us and them.

A final criticism of the analogy argument, also not very common but definitely compelling in light of what we know about mirror neurons, is the underestimation of the ability to access other minds. As we have seen, without resorting to any magic trick, our brains are capable of accessing other minds by using neural mechanisms of mirroring and simulation.

"Simulation"—I have used the word numerous times to describe what is going on in the brain of the observer of others' actions, and it is widely used in the field, but I am not entirely happy with it. To me, simulation implies some level of *conscious* effort, whereas a great deal of mirror neuron activity most likely reflects an experience-based, pre-reflective, and automatic form of understanding other minds. The father of phenomenology, Edmund Husserl, described the phenomenon (without referring to mirror neurons, of course) as "coupling." I like this term, though it may be too strong because it implies that the two individuals become just one entity. Recall that in chapter 5 our brain imaging data show how the sense of being the agent of one's own actions is maintained—in "spite" of mirror neurons—by enhancing the feedback we receive from our bodies. Recall also that in chapter 7 our single-unit recordings in neurological patients have revealed a special class of mirror neurons—super mirror neurons—that increase their firing rate for actions of the self, but decrease their firing rate for actions of other people. These two neural mechanisms allow us to internally represent self and other with some level of independence, even while they are mirroring each other.

The role of mirror neurons in intersubjectivity, then, may be more accurately described as allowing interdependence rather than pure "coupling." We have seen that, through mirror neurons, we can understand the intentions of others, thereby predicting—still in a pre-reflective way—their future behavior. The interdependence between self and other that mirror neurons allow shapes the social interactions between people, where the concrete encounter between self and other becomes the shared existential meaning that connects them deeply.

A NEW EXISTENTIALISM

In my lectures on mirror neurons I often conclude by saying that our research should be called existential neuroscience. I say this because the themes raised by mirror neuron research map well onto themes recurrent in existential phenomenology. The feedback from students and peers tends to be very positive on the phenomenology part of the equation, but much less on the existential part. I believe existentialism got a bad press at the peak of its popularity in the 1940s and 1950s, and it still gets a bad press now, probably because of the association with the ideas of dread and despair. The existential themes I am thinking about with mirror neuron research have nothing to do with dread and despair. If anything, they are optimistic and could be used to build a more empathetic, caring society.[3]

Obviously, phenomenology maps well onto mirror neuron research because only by "going back to the things themselves" could my friends in Parma have found these cells in the first place. Even theoreticians who more vocally claimed the closeness or intimacy between the self and others never proposed a natural phenomenon like mirror neurons. Interestingly, the only scientists who had even a glimmer of a mirror neuron system before it was discovered were those who do not typically theorize or passively observe, but rather those who build things. The roboticist Maja Matarić at USC told me that while struggling to build robots that could learn from experience and could imitate, she had thought of something similar to mirror neurons. Other roboticists also entertained such engineering "fantasies," which now turn out to be right on the mark.

Existentialism, on the other hand, invites us to embrace meaning in this world, the world of our experience, rather than identifying meaning on some metaphysical plane, outside of ourselves.[4] Mirror neurons are the cells in our brain that make our experience, mostly made of interactions with other people, deeply meaningful. This is why I call the mirror neuron research an existential neuroscience of sorts. This definition may sound like an oxymoron, since the dichotomy between analytic and continental (including existential) philosophy traditionally assigns detached hyperrational and scientific thinking to the analytic school and poetic, literary, or more generally artistic "culture" to the continental and existential school. However, there is one lesson we should have learned from mirror neurons by now: to be suspicious of rigid dichotomies (remember perception and action?).[5] The existentialists have constantly reminded us that what is worth understanding and knowing is our existence, the human condition, and that engagement and involvement are superior to a detached stance. Mirror neurons are brain cells that seem specialized in understanding our existential condition and our involvement with others. They show that we are not alone, but are biologically wired and evolutionarily designed to be deeply interconnected with one another.

There is also another existential theme that maps well onto mirror neuron properties. This theme goes back to the person who is considered the very first existential thinker,

Søren Kierkegaard. In *Fear and Trembling*, Kierkegaard proposed that our existence becomes meaningful only through our authentic commitment to the finite and temporal, a commitment that defines us. The neural resonance between self and other that mirror neurons allow is in my opinion the embodiment of such commitment. Our neurobiology—our mirror neurons—commits us to others. Mirror neurons show the deepest way we relate to and understand each other: they demonstrate that we are wired for empathy, which should inspire us to shape our society and make it a better place to live.

NEUROSCIENCE AND SOCIETY[6]

When we encounter each other, we share emotions and intentions. We are deeply interconnected at a basic, pre-reflective level. This we now know, and this *fact* seems to me a fundamental starting point for social behavior that has been largely neglected by an analytical tradition that emphasizes reflective behavior and differences among people. On the other hand, there is another fact staring us in the face: an atrocious world, literally—one filled with atrocities every day—and this despite a neurobiology wired for empathy and geared toward mirroring and sharing of meaning. Why is this?

I believe it is due to three main factors. First, we have seen in the phenomenon of imitative violence that the same neurobiological mechanisms facilitating empathy may produce, under specific circumstances and contexts, a behavior that is the opposite of an empathetic behavior. This is mostly a hypothesis right now, but it is a very strong one. If confirmed, this neuroscientific fact should inform policy making. Will it? I doubt it, for two reasons. First, our society is far from being ready to use scientific data to drive policy, especially in cases such as imitative violence, which involve an intricate relationship between financial interests and free speech. It's a tough policy issue with no easy answers, and I don't think it helps to confine science in general, and neuroscience in particular, to the ivory tower and the marketplace: discoveries are applied only to developing pharmacological treatments for neurological diseases, rarely to enhancing the well-being of the society as a whole. I would like to see at least an open discussion of the claim that neuroscientific discoveries could and should actually inform policy making. There is not much of that kind of thinking going on right now, and I believe we need it.

The second reason why there is resistance to the idea of neuroscience affecting policy has to do with the perceived threat to our notion of free will that is obviously linked to the argument on imitative violence. The research on mirror neurons implies that our sociality—perhaps the highest achievement of humans—is also a limiting factor of our autonomy as individuals. This is a major revision of long-standing beliefs. Traditionally, biological determinism of individual behavior is contrasted by a view of humans capable of rising above their biological makeup to define themselves through their ideas and their social codes. Mirror neuron research, however, suggests that our social codes are largely dictated by our biology. What should we do with this newly acquired knowledge?

Deny it because it is difficult to accept it? Or use it to inform policy and make our society a better one? I would obviously vote for the latter.

The second factor that has reduced the beneficial impact of our fundamental neurobiological drive to understand and to empathize is the "level" at which such a neurobiological drive works best. Mirror neurons are premotor neurons, remember, and thus are cells not really concerned with our reflective behavior. Indeed, mirroring behaviors such as the chameleon effect seem implicit, automatic, and pre-reflective. Meanwhile, society is obviously built on explicit, deliberate, reflective discourse. Implicit and explicit mental processes rarely interact; indeed, they can even dissociate. However, the neuroscientific discovery of mirror neurons has disclosed the pre-reflective neurobiological mechanisms of mirroring to our reflective level of understanding other people. I hope this book is welcomed into the fray. People do seem to have an intuitive understanding of how neural mechanisms for mirroring work. When told about this research, they catch on—at least in my experience. They can finally articulate what they already "knew" at a pre-reflective level. Indeed, the use in everyday language of the expression "being moved" by something is revealing of this pre-reflective level of understanding of the roots of empathy. People say that they are moved to sadness when they watch a tearjerker film; they are moved to joy when their child hits a home run and celebrates after rounding the bases. In some literal sense they are indeed moved. There is something like physical contact when they orchestrate movements in their mind when watching someone else. People seem to have the *intuition* that "being moved" is the basis of empathy and thus morality. My hope is that a more explicit level of understanding of our empathetic nature will at some point be a factor in the deliberate, reflective discourse that shapes society.

The third factor inhibiting what should be the positive impact of the mirroring network has to do with the powerful *local* effects of mirroring and imitation in shaping a variety of human cultures that are often not interconnected with one another and therefore end up clashing, as we often see these days, all around the world. In the existential phenomenology tradition, there is a strong emphasis on imitation of local traditions as a powerful shaper of the individual.[7] We become heirs of communal traditions. Who could doubt it? However, the powerful neurobiological mechanisms of mirroring that make this assimilation of local traditions possible could also *reveal* other cultures as long as such encounters are truly possible. Instead, we see exactly the opposite. True cross-cultural encounters are actually made impossible by the influence of massive belief systems—religious and political—that deny continuously the fundamental neurobiology that links us together.[8]

I believe we are at a point at which findings from neuroscience can significantly influence and change our society and our understanding of ourselves. It is high time we consider this option seriously. Our knowledge of the powerful neurobiological mechanisms underlying human sociality provides an invaluable resource for helping us determine how to reduce violent behavior, increase empathy, and open ourselves to other cultures without forgetting our own. We have evolved to connect deeply with other human beings. Our awareness of this fact can and should bring us even closer to one another.

Afterword

I am writing this afterword on Inauguration Day. These are days of high hope, the hope that Barack Obama's campaign and presidential election have so brilliantly raised in all of us. Amazingly, in his first speech as President, Obama has said that he wants to "restore science to its rightful place." Obviously, I couldn't agree more. *Mirroring People* also ends on a hopeful note, the hope that science and scientific thinking may play an important role in our society.

The purpose of this afterword is to provide a brief update on the scientific progress and research on mirror neurons from the past year. Science obviously never stops; every month the results of new experiments are announced at scientific conferences and new scientific papers are published in specialized journals. A book, however, is written over a long period of time, and the abundance of new data, new findings, and new interpretations that had emerged during the writing of *Mirror-*

ing People presented a challenge to me. I asked myself many times as the manuscript entered the final stages of editing: "Should I include these new results at the cost of altering the structure of the argument, or the rhythm of the book?" Luckily, I am no longer in that rather uncomfortable situation, and this afterword, written in a compressed time frame, is only intended to serve as a snapshot of the most important research findings that have appeared since I finished the book.

There have been so many different studies that I cannot possibly summarize them all, which perhaps is testimony to the fact that there is a lot of excitement in the neurosciences about mirror neurons. Indeed, research on neural mirroring has begun to expand beyond primates. Two recent scientific articles published in the highly regarded journal *Nature* have reported the existence of auditory-vocal mirror neurons in songbirds, suggesting that these cells may be critical for songbird learning. I expect many more studies on mirror neurons in non-primates in the future, and that these studies will be extremely important to understanding the role of mirror neurons in animal behavior. In this afterword, however, I will focus on studies in monkeys and humans, and in particular will highlight research trends that illustrate our understanding of mirror neurons and our social brain.

There are three major trends in mirror neuron research today. One includes studies that investigate what we scientists call "individual differences." Some pioneering studies of this kind have been already discussed in the book, specifically those that investigated the reduced activity of the mirror neuron system in patients with autism, and how it correlates with the severity of autism; or the studies that looked at mirror neuron activity among adolescents or healthy adults and correlated it with a variety of measures of social competence and empathy. Many studies of this kind have been presented at conferences recently. They correlate a variety of personality and psychological measures with activity in mirror neuron areas. The more we learn about the system in our brain, the more it seems a cornerstone of our ability to socialize with others. It is too early to draw strong conclusions or to summarize many of the findings in a few sentences, but there is definitely a sense that we may soon reach an amazing understanding of the biological foundations of our most complex behavior—that is, our social behavior.

The second trend relates to the development of the mirror neuron system. It is extremely important to understand how mirror neurons are formed early on in life, and which factors favor a healthy development of these cells. Many colleagues share my belief, and they are trying to investigate the mirror neuron system in the infant brain. Unfortunately, for many technical and ethical reasons, it is difficult to conduct these kinds of experiments. The most promising study on the mirror neurons system in the infant brain is currently being performed in the lab of Steve Suomi at NIH. The scientist leading this study is Pier Francesco Ferrari, the ethologist and neurophysiologist who led the experiment on mirror neurons and tool-use described in the first chapter of this book. Ferrari is studying baby monkeys from their very first few days of life, using EEG and behavioral observations. At this time, Ferrari, Suomi, and their colleagues have not yet published their findings. Ferrari, however, presented some preliminary observations at two small conferences I attended in the last few months. Two of his findings are particularly intriguing. First, Ferrari and his colleagues looked at the ability of the infant monkey to imitate. In this experiment, the baby monkeys were prevalently imitating relatively simple facial expressions, for instance tongue protrusion, as in the study on human infants led by Meltzoff that I described in the second chapter of the book. Ferrari and his colleagues found that some infant monkeys are good imitators, whereas others do not imitate well at all. This is not really surprising. Monkeys, and humans too, of course, all differ from one another to some degree. The fundamental question, however, is whether being a good imitator very early in life facilitates learning. The scientists found that the good imitators developed the ability to grasp objects earlier and better than the bad imitators. This result confirms not only how important imitative learning is early in life, but also how hand and mouth are tightly coupled in the brain, as discussed in chapter 3.

The second important finding from Ferrari's experiment on baby monkeys is related to the EEG data. Remember that one of the "biomarkers" of mirror neuron activity we have discussed is the suppression of the mu rhythm. In the infant monkey brain, Ferrari and colleagues have found something similar, that is, the suppression of slow 3-5 Hz rhythm around motor areas while the baby monkey is observing the actions of somebody else. It is still too early to have a full understanding of this phenomenon; however, if we assume that it represents yet another biomarker of mirror neuron activity, we have to conclude that the infant brain already possesses some mirror neurons.

The third and final trend of recent studies on mirror neurons is related to developing a more refined description of their properties. Two recent findings are extremely exciting, and both remind us of the themes and ideas developed by our old friend, the French phenomenologist Maurice Merleau-Ponty. They are related to the concepts of space and goals. Remember that area F5, the brain area where mirror neurons were originally discovered, borders brain area F4, where space is coded in two major maps, one for peripersonal space (the space that surrounds the body, where we can reach for objects and grab them), and the other one for extrapersonal space (the space we can't reach). New data from Leo Fogassi, one of the researchers who originally discovered mirror neurons, show that some mirror neurons change their responses to observed action on the basis of *where the action is located*. The old studies on mirror neurons were all performed by showing the actions far away from the monkey (that is, all the actions were in extrapersonal space). This was done to avoid the criticism that the firing of the cell—while the monkey was simply observing—was due to some form of *motor preparation*, getting ready to act because somebody else was grasping things close to the observing monkey. In the new study, Fogassi first performed some grasping actions in the peripersonal space of the monkey. Not surprisingly, mirror neurons fired at the sight of the action. Then he had the clever idea of placing a glass screen between himself and the monkey. Now the monkey can still see the action (performed in her peripersonal space) but cannot possibly act on the object (because the screen gets in the way). In this situation, some mirror neurons no longer fire. These cells apparently respond to the actions of other people in an *operational* way. If the observer cannot intervene on the observed action, there is no firing. These neurons are all about engagement, being involved in social interaction in a fundamentally active way. If we can't intervene, these cells no longer care about what we see.

The most compelling evidence in support of the idea that mirror neurons (and motor neurons in general) are less concerned with specific details of the action and more with its goal, comes from another study from the group of Rizzolatti (Umiltà et al., "When pliers become fingers in the monkey motor system." PNAS 105:2209-2213; 2008) that trained monkeys to use reverse pliers to grab small objects like peanuts and raisins. With regular pliers, the monkey had to *flex* the fingers to grab the object (closing the hand). With reverse pliers, the monkey had to *extend* the fingers to grab the object (opening the hand). The movements performed by the monkey were exactly the *opposite* of each other. Amazingly, however, the same grasping cells in area F5 that fired when the monkey flexed the fingers while using the regular pliers also fired when the monkey extended the fingers while using the reverse pliers to grab the objects. Mirror neurons obviously fired when the monkey simply observed somebody else using regular and reverse pliers. This beautiful study confirmed in the most compelling way that both motor cells in area F5 and mirror neurons are concerned more with the goal we try to achieve with our own actions than with the motor details of how to achieve our goals. This concept has obviously profound implications for motor learning and rehabilitation, but also reveals an important aspect of how the motor system and mirror neurons code information. This coding is not *shallow*; it does not stop at the superficial elements of our own and other people's actions. It is a *deep* coding, a coding that allows us to deeply understand other people's goals.

The more we investigate the properties of mirror neurons, the more we understand how these cells help us to be empathic and fundamentally attuned to other people. This is perhaps the most important finding of all, and it is a beautiful one.

Notes

One: Monkey See, Monkey Do

1. V. S. Ramachandran, "Mirror Neurons and Imitation Learning as the Driving Force Behind 'the Great Leap Forward' in Human Evolution," *Edge* 69, June 29, 2000 (www.edge.org/3rd_culture/ramachandran/ramachandran_index.html). Notes such as this one will be used for references and to make comments that may be of interest mainly to specialists.

2. Truth to be told, Rizzolatti and his colleagues were definitely more open-minded than the average neuroscientist and more "ready" for the new discovery. This is probably why they discovered mirror neurons. Those same phenomena, occurring under the nose of more narrow-minded neuroscientists, would have gone unnoticed. Who knows how many times the firing of mirror neurons did go unnoticed in neurophysiology labs!

3. Gentilucci, M., L. Fogassi, G. Luppino, et al., "Functional Organization of Inferior Area 6 in the Macaque Monkey. I. Somatotopy and the Control of Proximal Movements," *Experimental Brain Research* 71 (1988):475–90; Rizzolatti, G., R. Camarda, L. Fogassi, et al., "Functional Organization of Inferior Area 6 in the Macaque Monkey. II. Area F5 and the Control of Distal Movements," *Experimental Brain Research* 71 (1998):491–507.

4. Rizzolatti, G., C. Scandolara, M. Matelli, and M. Gentilucci, "Afferent

Properties of Periarcuate Neurons in Macaque Monkeys. II. Visual Responses," *Behavioural Brain Research* 2 (1981):147–63; Rizzolatti, G., C. Scandolara, M. Matelli, and M. Gentilucci, "Afferent Properties of Periarcuate Neurons in Macaque Monkeys. I. Somatosensory Responses," *Behavioural Brain Research* 2 (1981):125–46.

5. Gallese, V., and A. Goldman, "Mirror Neurons and the Simulation Theory of Mind-reading," *Trends in Cognitive Sciences* 2 (1998):493–501.

6. Rizzolatti, G., L. Riggio, I. Dascola, and C. Umiltà, "Reorienting Attention Across the Horizontal and Vertical Meridians: Evidence in Favor of a Premotor Theory of Attention," *Neuropsychologia* 25 (1987):31–40; Corbetta, M., E. Akbudak, T. E. Conturo, et al., "A Common Network of Functional Areas for Attention and Eye Movements," *Neuron* 21 (1998):761–73.

7. Rizzolatti, G., et al., "Functional Organization of Inferior Area 6 in the Macaque Monkey. II," 491–507.

8. Gallese, V., L. Fadiga, L. Fogassi, and G. Rizzolatti, "Action Recognition in the Premotor Cortex," *Brain* 119 (Pt. 2) (1996):593–609; Rizzolatti, G., and L. Craighero, "The Mirror-Neuron System." *Annual Review of Neuroscience* 27 (2004):169–92.

9. Di Pellegrino, G., L. Fadiga, L. Fogassi, et al., "Understanding Motor Events: A Neurophysiological Study," *Experimental Brain Research* 91 (1992):176–80.

10. Arbib, M. A., "From Monkeylike Action Recognition to Human Language: An Evolutionary Framework for Neurolinguistics," *Behavioral and Brain Science* 28 (2005):105–24; discussion 125–67.

11. Ferrari, P. F., V. Gallese, G. Rizzolatti, and L. Fogassi, "Mirror Neurons Responding to the Observation of Ingestive and Communicative Mouth Actions in the Monkey Ventral Premotor Cortex," *European Journal of Neuroscience* 17 (2003):1703–14.

12. Umiltà, M. A., E. Kohler, V. Gallese, et al., "I Know What You Are Doing: A Neurophysiological Study," *Neuron* 31 (2001):155–65.

13. Fogassi, L., P. F. Ferrari, B. Gesierich, et al., "Parietal Lobe: From Action Organization to Intention Understanding," *Science* 308 (2005):662–67.

14. Kohler, E., C. Keysers, M. A. Umiltà, et al., "Hearing Sounds, Understanding Actions: Action Representation in Mirror Neurons," *Science* 297

(2002):846–48; Keysers, C., E. Kohler, M. A. Umiltà, et al., "Audiovisual Mirror Neurons and Action Recognition." *Experimental Brain Research* 153 (2003):628–36.

15. Rizzolatti, G., and M. A. Arbib, "Language Within Our Grasp," *Trends in Neuroscience* 21 (1998):188–94.

16. Liberman, A. M., and I. G. Mattingly, "The Motor Theory of Speech Perception Revised," *Cognition* 21 (1985):1–36.

17. Whiten, A., J. Goodall, W. C. McGrew, et al., "Cultures in Chimpanzees," *Nature* 399 (1999):682–85.

18. Ferrari, P. F., S. Rozzi, and L. Fogassi, "Mirror Neurons Responding to Observation of Actions Made with Tools in Monkey Ventral Premotor Cortex," *Journal of Cognitive Neuroscience* 17 (2005):212–26.

19. Romanes, G. J., *Mental Evolution in Animals* (London: Kegan Paul Trench & Co., 1883); Hurley, S., and N. Chater, *Perspectives on Imitation: From Neuroscience to Social Science* (Cambridge, MA: MIT Press, 2005).

20. Ferrari, P. F., E. Visalberghi, A. Paukner, et al., "Neonatal Imitation in Rhesus Macaques," *PLoS Biology* 4 (2006):e302.

21. Voelkl, B., and L. Huber, "True Imitation in Marmosets," *Animal Behavior* 60 (2000):195–202.

22. Paukner, A., J. R. Anderson, E. Borelli, et al., "Macaques (*Macaca nemestrina*) Recognize When They Are Being Imitated," *Biology Letters* 1 (2005):219–22.

Two: Simon Says

1. I found this quote in the charming book on emotional contagion by E. Hatfield, J. T. Cacioppo, and R. L. Rapson, *Emotional Contagion* (New York: Cambridge University Press, 1994). However, the book does not specify the original source.

2. Meltzoff, A. N., and M. K. Moore, "Imitation of Facial and Manual Gestures by Human Neonates," *Science* 198 (1977):74–78; Piaget, J., *Play, Dreams and Imitation in Childhood* (London: Routledge, 1951).

3. Nadel, J. "Imitation and Imitation Recognition: Functional Use in Preverbal Infants and Nonverbal Children with Autism," in A. N. Meltzoff and W. Prinz, *The Imitative Mind: Development, Evolution, and Brain Bases* (Cambridge, UK: Cambridge University Press, 2002); Eckerman, C. O.,

and S. M. Didow, "Nonverbal Imitation and Toddlers' Mastery of Verbal Means of Achieving Coordinated Actions," *Developmental Psychology* 32 (1996):141–52.

4. Dawkins, R., *The Selfish Gene* (Oxford, UK: Oxford University Press, 1976); Blackmore, S., *The Meme Machine* (Oxford, UK: Oxford University Press, 1999).

5. See, for instance, Dennett, D., *Consciousness Explained* (Boston: Little, Brown, 1991); Hull, D. L., "The Naked Meme," in H. C. Plotkin, ed., *Learning Development and Culture: Essays in Evolutionary Epistemology* (London: Wiley, 1982).

6. Berger, S. M., and S. W. Hadley, "Some Effects of a Model's Performance on an Observer's Electromyographic Activity," *American Journal of Psychology* 88 (1975):263–76.

7. Rizzolatti, G., L. Fadiga, M. Matelli, et al., "Localization of Grasp Representations in Humans by PET: 1. Observation Versus Execution," *Experimental Brain Research* 111 (1996):246–52; Grafton, S. T., M. A. Arbib, L. Fadiga, and G. Rizzolatti, "Localization of Grasp Representations in Humans by Positron Emission Tomography. 2. Observation Compared with Imagination," *Experimental Brain Research* 112 (1996):103–11.

8. Fadiga, L., L. Fogassi, G. Pavesi, and G. Rizzolatti, "Motor Facilitation During Action Observation: A Magnetic Stimulation Study," *Journal of Neurophysiology* 73 (1995):2608–11.

9. Prinz, W., "An Ideomotor Approach to Imitation," in S. Hurley, and N. Chater, *Perspectives on Imitation: From Neuroscience to Social Science. Volume 1: Mechanisms of Imitation and Imitation in Animals* (Cambridge, MA: MIT Press, 2005), 141–56.

10. James, W., *Principles of Psychology* (New York: Holt, 1890).

11. Gleissner, B., A. N. Meltzoff, and H. Bekkering, "Children's Coding of Human Action: Cognitive Factors Influencing Imitation in Three-Year-Olds," *Developmental Science* 3 (2000):405–14; Bekkering, H., A. Wohlschläger, and M. Gattis, "Imitation of Gestures in Children Is Goal-Directed," *Quarterly Journal of Experimental Psychology* A 53 (2000):153–64; Wohlschläger, A., and H. Bekkering, "Is Human Imitation Based on a Mirror-Neurone System? Some Behavioural Evidence," *Experimental Brain Research* 143 (2002):335–41.

12. Koski, L., A. Wohlschläger, H. Bekkering, et al., "Modulation of Motor and Premotor Activity During Imitation of Target-Directed Actions," *Cerebral Cortex* 12 (2002):847–55.

13. Wapner, S., and L. Cirillo, "Imitation of a Model's Hand Movement: Age Changes in Transposition of Left-Right Relations." *Child Development* 39 (1968):887–94; Koski, L., M. Iacoboni, M. C. Dubeau, et al., "Modulation of Cortical Activity During Different Imitative Behaviors," *Journal of Neurophysiology* 89 (2003):460–71.

14. Hatfield, et al., *Emotional Contagion*; Bavelas, J. B., A. Black, N. Chovil, et al., "Form and Function in Motor Mimicry: Topographic Evidence That the Primary Function Is Communication," *Human Communications Research* 14 (1988):275–99; LaFrance, M., "Posture Mirroring and Rapport," in M. Davis, ed., *Interaction Rhythms: Periodicity in Communicative Behavior* (New York: Human Sciences Press, 1982), 279–98.

15. Rogers, S. J., and B. F. Pennington, "A Theoretical Approach to the Deficits in Infantile Autism," *Developmental Psychology* 3 (1991): 137–62; Whiten, A., and J. D. Brown, "Imitation and the Reading of Other Minds: Perspectives from the Study of Autism, Normal Children and Non-human Primates," in S. Braten, ed., *Intersubjective Communication and Emotion in Early Ontogeny* (Cambridge, UK: Cambridge University Press, 1999), 260–80; Williams, J. H., A. Whiten, T. Suddendorf, et al., "Imitation, Mirror Neurons and Autism," *Neuroscience and Biobehavioral Reviews* 25 (2001):287–95.

16. Gallese, V., and A. Goldman, "Mirror Neurons and the Simulation Theory of Mind-reading," *Trends in Cognitive Science* 2 (1998):493–501; Carruthers, P., and P. Smith, *Theories of Theories of Mind* (Cambridge, UK: Cambridge University Press, 1996); Goldman, A. I., "Imitation, Mind Reading, and Simulation," in S. Hurley and N. Chater, eds., *Perspectives on Imitation: From Neuroscience to Social Science, Volume 2: Imitation, Human Development, and Culture* (Cambridge, MA: MIT Press, 2005), 79–94; Gordon, R. M., "Intentional Agents Like Myself," in Hurley and Chater, *Perspectives on Imitation, Volume 2*, 95–106; Goldman, A., *Simulating Minds: The Philosophy, Psychology, and Neuroscience of Mindreading* (New York: Oxford University Press, 2006).

17. Iacoboni, M., I. Molnar-Szakacs, V. Gallese, et al., "Grasping the Inten-

tions of Others with One's Own Mirror Neuron System," *PLoS Biology* 3 (2005):e79.

18. Gallese, V., "Intentional Attunement: A Neurophysiological Perspective on Social Cognition and Its Disruption in Autism," *Brain Research* 1079 (2006):15–24, Merleau-Ponty, M., *Phenomenology of Perception* (London: Routledge, 1945).

Three: Grasping Language

1. Napier, J., *Hands* (New York: Pantheon Books, 1980).

2. McNeill, D., *Hand and Mind: What Gestures Reveal About Thought* (University of Chicago Press, 1992).

3. Goldin-Meadow, S., "When Gestures and Words Speak Differently," *Current Directions in Psychological Science* 6 (1997):138–43; Goldin-Meadow, S., "The Role of Gesture in Communication and Thinking," *Trends in Cognitive Sciences* 3 (1999):419–29.

4. Alibali, M. W., D. C. Heath, and H. J. Myers, "Effects of Visibility Between Speaker and Listener on Gesture Production: Some Gestures Are Meant to Be Seen," *Journal of Memory and Language* 44 (2001): 169–88.

5. Molnar-Szakacs, I., S. M. Wilson, and M. Iacoboni, "I See What You Are Saying: The Neural Correlates of Gesture Perception," Program No. 128.7. *2005 Abstract Viewer, CD-ROM*. Washington, DC: Society for Neuroscience meeting.

6. Rizzolatti, G., and M. A. Arbib, "Language Within Our Grasp," *Trends in Neuroscience* 21 (1998):188–94; G. von Bonin and P. Bailey, *The Neocortex of Macaca Mulatta* (Urbana: University of Illinois Press, 1947).

7. Iverson, J. M., and E. Thelen, "Hand, Mouth and Brain. The Dynamic Emergence of Speech and Gesture," *Journal of Consciousness Studies* 6 (1999):19–40; Goldin-Meadow, "The Role of Gesture in Communication and Thinking," 419–29.

8. Greenfield, P. M., "Language, Tools and Brain: The Ontogeny and Phylogeny of Hierarchically Organized Sequential Behavior," *Behavioral and Brain Sciences* 14 (1991):531–95; Molnar-Szakacs, I., J. Kaplan, P. M. Greenfield, and M. Iacoboni, "Observing Complex Action Sequences: The Role of the Frontoparietal Mirror Neuron System," *Neuroimage* 33

(2006):923–35; Greenfield, P., "Implications of Mirror Neurons for the Ontogeny and Phylogeny of Cultural Processes: The Examples of Tools and Language," in M. A. Arbib, ed., *Action to Language Via the Mirror Neuron System* (New York: Cambridge University Press, 2006), 503–35.

9. Heiser, M., M. Iacoboni, F. Maeda, et al., "The Essential Role of Broca's Area in Imitation," *European Journal of Neuroscience* 17 (2003):1123–28.

10. Glenberg, A. M., and M. P. Kaschak, "Grounding Language in Action," *Psychonomic Bulletin and Review* 9 (2002):558–65.

11. Ochs, E., P. Gonzales, and S. Jacoby, " 'When I Come Down I'm in the Domain State': Grammar and Graphic Representation in the Interpretive Activity of Physicists," in E. Ochs, E. A. Schegloff, and S. A. Thompson, eds., *Interaction and Grammar* (New York: Cambridge University Press, 1996), 328–69.

12. Gallese, V., and G. Lakoff, "The Brain's Concepts: The Role of the Sensory-Motor System in Conceptual Knowledge," *Cognitive Neuropsychology* 22 (2005):455–79.

13. Aziz-Zadeh, L., S. M. Wilson, G. Rizzolatti, and M. Iacoboni, "Congruent Embodied Representations for Visually Presented Actions and Linguistic Phrases Describing Actions," *Current Biology* 16 (2006):1818–23.

14. Garrod, S., and M. J. Pickering, "Why Is Conversation So Easy?" *Trends in Cognitive Sciences* 8 (2004):8–11.

15. Brennan, S. E., and H. H. Clark, "Conceptual Pacts and Lexical Choice in Conversation," *Journal of Experimental Psychology: Learning, Memory, and Cognition* 22 (1996):1482–93; Schober, M. F., and H. H. Clark, "Understanding by Addressees and Over-Hearers," *Cognitive Psychology* 21 (1989):211–32.

16. Goodwin, C., "Restarts, Pauses, and the Achievement of a State of Mutual Gaze at Turn-beginning," *Sociological Inquiry* 50 (1980):272–302; Kendon, A., "Some Functions of Gaze-direction in Social Interaction," *Acta Psychologica* 26 (1967):22–63; Goodwin, C., and J. Heritage, "Conversation Analysis," *Annual Review of Anthropology* 19 (1990):283–307.

17. Kegl, J., "The Nicaraguan Sign Language Project: An Overview," *Signpost* 7 (1994):24–31.

18. See, for instance, S. Pinker, *The Language Instinct* (New York: Morrow, 1994).

19. Tomasello, M., "The Item-based Nature of Children's Early Syntactic Development," *Trends in Cognitive Sciences* 4 (2000):156–63.

20. Clark, H., *Using Language* (Cambridge, UK: Cambridge University Press, 1996); Garrod, S., and A. Anderson, "Saying What You Mean in Dialogue: A Study in Conceptual and Semantic Co-ordination," *Cognition* 27 (1987):181–218; Galantucci, B., "An Experimental Study of the Emergence of Human Communication Systems," *Cognitive Science* 29 (2005):737–67.

21. Aziz-Zadeh, L., M. Iacoboni, E. Zaidel, et al., "Left Hemisphere Motor Facilitation in Response to Manual Action Sounds," *European Journal of Neuroscience* 19 (2004):2609–12; Gazzola, V., L. Aziz-Zadeh, and C. Keysers, "Empathy and the Somatotopic Auditory Mirror System in Humans," *Current Biology* 16 (2006):1824–29.

22. McGurk, H., and J. MacDonald, "Hearing Lips and Seeing Voices," *Nature* 264 (1976):746–48.

23. Liberman, A. M., and I. G. Mattingly, "The Motor Theory of Speech Perception Revised," *Cognition* 21 (1985):1–36.

24. Fadiga, L., L. Craighero, G. Buccino, and G. Rizzolatti, "Speech Listening Specifically Modulates the Excitability of Tongue Muscles: A TMS Study," *European Journal of Neuroscience* 15 (2002):399–402.

25. Wilson, S. M., A. P. Saygin, M. I. Sereno, and M. Iacoboni, "Listening to Speech Activates Motor Areas Involved in Speech Production," *Nature Neuroscience* 7 (2004):701–702.

26. Meister, I., S. M. Wilson, C. Deblieck, et al., "The Essential Role of Premotor Cortex in Speech Perception," *Current Biology* 17 (2007):1692–96.

27. Warren, J. E., D. A. Sauter, F. Eisner, et al., "Positive Emotions Preferentially Engage an Auditory-Motor 'Mirror' System," *Journal of Neuroscience* 26 (2006):13067–75.

Four: See Me, Feel Me

1. Smith, A., *The Theory of Moral Sentiments* (Oxford, UK: Clarendon Press, 1976).

2. Gallese, V., "The 'Shared Manifold' Hypothesis," *Journal of Consciousness Studies* 8 (2001):33–50; Lipps, T., "*Einfühlung, innere nachahmung und*

NOTES TO PAGES 110—125

organ-enempfindung," in *Archiv für die Gesamte Psychologie*, volume I, part 2 (Leipzig: W. Engelmann, 1903).

3. Dimberg, U., "Facial Reactions to Facial Expressions," *Psychophysiology* 19 (1982):643–47.

4. Hatfield et al., *Emotional Contagion*.

5. Niedenthal, P. M., L. W. Barsalou, P. Winkielman, et al., "Embodiment in Attitudes, Social Perception, and Emotion," *Personality and Social Psychology Reviews* 9 (2005):184–211.

6. Chartrand, T. L., and J. A. Bargh, "The Chameleon Effect: The Perception-Behavior Link and Social Interaction," *Journal of Personality & Social Psychology* 76 (1999):893–910.

7. Zajonc, R. B., P. K. Adelmann, S. T. Murphy, et al., "Convergence in the Physical Appearance of Spouses," *Motivation and Emotion* 11 (1987):335–46; Cole, J., "Empathy Needs a Face," *Journal of Consciousness Studies* 8 (2001):51–68; Merleau-Ponty, M., *The Primacy of Perception* (Evanston, IL: Northwestern University Press, 1964).

8. Augustine, J. R., "Circuitry and Functional Aspects of the Insular Lobes in Primates Including Humans," *Brain Research Reviews* 22 (1996):229–94.

9. Poe, E. A., *The Tell-Tale Heart and Other Writings* (New York: Bantam Books, 1982); Darwin, C., *The Expression of the Emotions in Man and Animals* (University of Chicago Press, 1965); James, W. (1890), "What Is an Emotion?" in C. Calhoun and R. C. Solomon, eds., *What Is an Emotion?* (New York: Oxford University Press, 1984), 125–42.

10. Carr, L., M. Iacoboni, M. C. Dubeau, et al., "Neural Mechanisms of Empathy in Humans: A Relay from Neural Systems for Imitation to Limbic Areas," *Proceedings of the National Academy of Sciences USA* 100 (2003):5497–5502.

11. Hutchison, W. D., K. D. Davis, A. M. Lozano, et al., "Pain-related Neurons in the Human Cingulate Cortex," *Nature Neuroscience* 2 (1999):403–405.

12. Avenanti, A., D. Bueti, G. Galati, et al., "Transcranial Magnetic Stimulation Highlights the Sensorimotor Side of Empathy for Pain," *Nature Neuroscience* 8 (2005):955–60.

13. Singer, T., B. Seymour, J. O'Doherty, et al., "Empathy for Pain Involves

the Affective but Not Sensory Components of Pain," *Science* 303 (2004):1157–62.

14. Antonio Damasio's "as if loop" theory, although not directly referring to mirror neurons, at least in its original versions that preceded the discovery of mirror neurons, also invoked a central role of simulative processes in emotion. See Damasio, A. R., *Descartes' Error: Emotion, Reason, and the Human Brain* (New York: Putnam, 1994); Damasio, A. R., *The Feeling of What Happens: Body and Emotion in the Making of Consciousness* (New York: Harcourt Brace, 1999); Damasio, A. R., *Looking for Spinoza: Joy, Sorrow, and the Feeling Brain* (Orlando, FL: Harcourt, 2003).

15. Haviland, J. M., and M. Lilac, "The Induced Affect Response: 10-Week-Old Infants' Responses to Three Emotion Expressions," *Developmental Psychology* 23 (1987):97–104; Termine, N. T., and C. E. Izard, "Infants' Response to Their Mother's Expressions of Joy and Sadness," *Developmental Psychology* 24 (1988):223–29.

16. Bernieri, F. J., J. S. Reznick, and R. Rosenthal, "Synchrony, Pseudosynchrony, and Dissynchrony: Measuring the Entrainment Process in Mother-Infant Interactions," *Journal of Personality and Social Psychology* 54 (1988):243–53.

17. Rizzolatti, G., and G. Luppino, "The Cortical Motor System," *Neuron* 31 (2001):889–901.

Five: Facing Yourself

1. Iacoboni, M., R. P. Woods, M. Brass, et al., "Cortical Mechanisms of Human Imitation," *Science* 286 (1999):2526–28.

2. Zahavi, D., "Beyond Empathy: Phenomenological Approaches to Intersubjectivity," *Journal of Consciousness Studies* 8 (2001):151–67.

3. Asendorpf, J. B., and P.-M. Baudonniere, "Self-awareness and Other-awareness: Mirror Self-recognition and Synchronic Imitation Among Unfamiliar Peers," *Developmental Psychology* 29 (1993):88–95.

4. Keenan, J. P., G. G. Gallup, and D. Falk, *The Face in the Mirror: The Search for the Origins of Consciousness* (New York: Ecco, 2003).

5. Gallup, G. G., "Chimpanzees: Self-recognition," *Science* 167 (1970): 86–87.

6. Miles, H., "Me Chantek: The Development of Self-Awareness in a Sign-

ing Orangutan," in S. Parker and R. Mitchell, *Self-awareness in Animals and Humans: Developmental Perspectives* (Cambridge, UK: Cambridge University Press, 1994), 254–72.

7. Reiss, D., and L. Marino, "Mirror Self-recognition in the Bottlenose Dolphin: A Case of Cognitive Convergence," *Proceedings of the National Academy of Sciences USA* 98 (2001):5937–42; Rendell, L., and H. Whitehead, "Culture in Whales and Dolphins," *Behavioral and Brain Sciences* 24 (2001):309–24; discussion 324–82.

8. Gallup, G. G., "Self-awareness and the Emergence of Mind in Primates," *American Journal of Primatology* (1982):237–48; Povinelli, D. J., "Failure to Find Self-recognition in Asian Elephants (*Elephas maximus*) in Contrast to Their Use of Mirror Cues to Discover Hidden Food," *Journal of Comparative Psychology* 103 (1989):122–31; Plotnick, J. M., F.B.M. de Waal, and D. Reiss, "Self-Recognition in an Asian Elephant," *Proceedings of the National Academy of Sciences USA* 103 (2006):17053–57.

9. Amsterdam, B., "Mirror Self-image Reactions Before Age Two," *Developmental Psychobiology* 5(1972):297–305.

10. Sperry, R. W., E. Zaidel, and D. Zaidel, "Self-recognition and Social Awareness in the Deconnected Minor Hemisphere," *Neuropsychologia* 17 (1979):153–66.

11. Uddin, L. Q., J. Rayman, and E. Zaidel, "Split-brain Reveals Separate but Equal Self-recognition in the Two Cerebral Hemispheres," *Consciousness and Cognition* 14 (2005):633–40.

12. Kourtzi, J., and N. Kanwisher, "Activation in Human MT/MST by Static Images with Implied Motion," *Journal of Cognitive Neuroscience* 12 (2000):48–55; Urgesi, C., V. Moro, M. Candidi, et al., "Mapping Implied Body Actions in the Human Motor System," *Journal of Neuroscience* 26 (2006):7942–49.

13. Uddin, L. Q., J. T. Kaplan, I. Molnar-Szakacs, et al., "Self-face Recognition Activates a Frontoparietal 'Mirror' Network in the Right Hemisphere: An Event-related fMRI Study," *Neuroimage* 25 (2005): 926–35.

14. Feinberg, T., and R. Shapiro, "Misidentification-Reduplication and the Right Hemisphere," *Neuropsychiatry, Neuropsychology, and Behavioral Neurology* 2 (1989):39–48; Spangenberger, K., M. Wagner, and D. Bachman,

"Neuropsychological Analysis of a Case of Abrupt Onset Following a Hypotensive Crisis in a Patient with Vascular Dementia," *NeuroCase* 4 (1998):149–54; Breen, N., D. Caine, and M. Coltheart, "Mirrored-Self Misidentification: Two Cases of Focal Onset Dementia," *NeuroCase* 7 (2001):239–54.

15. This type of brain stimulation is obviously entirely safe for the subjects.

16. Uddin, L., I. Molnar-Szakacs, E. Zaidel, et al., "rTMS to the Right Inferior Parietal Area Disrupts Self-Other Discrimination," *Social Cognitive and Affective Neuroscience* 1 (2006):65–71.

17. Feinberg, T. E., L. D. Haber, and N. E. Leeds, "Verbal Asomatognosia," *Neurology* 40 (1990):1391–94.

18. Trevarthen, C., "Communication and Cooperation in Early Infancy: A Description of Primary Intersubjectivity," in M. Bullowa, ed., *Before Speech* (New York: Cambridge University Press, 1979).

Six: Broken Mirrors

1. Shimada, S., and K. Hiraki, "Infant's Brain Responses to Live and Televised Action," *Neuroimage* 32 (2006):930–39.

2. Flanagan, J. R., and R. S. Johansson, "Action Plans Used in Action Observation," *Nature* 424 (2003):769–71.

3. Falck-Ytter, T., G. Gredebäck, and C. von Hofsten, "Infants Predict Other People's Action Goals," *Nature Neuroscience* 9 (2006):878–79.

4. Hari, R., N. Forss, S. Avikainen, et al., "Activation of Human Primary Motor Cortex During Action Observation: A Neuromagnetic Study," *Proceedings of the National Academy of Sciences USA* 95 (1998):15061–65.

5. Davis, M. H., "Measuring Individual Differences in Empathy: Evidence for a Multidimensional Approach," *Journal of Personality & Social Psychology* 44 (1983):113–26; Cairns, R. B., M-C. Leung, S. D. Gest, et al., "A Brief Method for Assessing Social Development: Structure, Reliability, Stability, and Developmental Validity of the Interpersonal Competence Scale," *Behaviour Research and Therapy* 33 (1995):725–36.

6. Pfeifer, J., M. Iacoboni, J. C. Mazziotta, and M. Dapretto, "Mirroring Others' Emotions Relates to Empathy and Interpersonal Competence in Children," *NeuroImage* (2008), in press.

7. Ritvo, S., and S. Provence, "From Perception and Imitation in Some

tion," *The Psychoanalytic Study of the Child, Volume VIII* (New York: International Universities Press, 1953), 155–61. I have to thank Ami Klin at Yale University for sending me this article.

8. Gopnik, A., A. N. Meltzoff, and P. K. Kuhl, *The Scientist in the Crib* (New York: Perennial, 2001). Meltzoff, one of the authors of this book, has recently modified his position. His "like me" hypothesis of social cognition is reminiscent of simulation theory. See his recent papers: Meltzoff, A. N., "Imitation and Other Minds: the 'Like Me' Hypothesis," in Hurley and Chater, *Perspectives on Imitation, Volume 2*, 55–77; Meltzoff, A. N., " 'Like Me': A Foundation for Social Cognition," *Developmental Science* 10:126–34; Meltzoff, A. N., " 'Like Me' Framework for Recognizing and Becoming an Intentional Agent," *Acta Psychologica* 124 (2007):26–43.

9. Rogers, S. J., and B. F. Pennington, "A Theoretical Approach to the Deficits of Infantile Autism," *Development & Psychopathology* 3 (1991):137–62.

10. Hobson, P., *The Cradle of Thought* (London: Pan Macmillan, 2002); Weeks, S. J., and R. P. Hobson, "The Salience of Facial Expression for Autistic Children," *Journal of Child Psychology and Psychiatry* 28 (1987):137–52.

11. Hobson, *The Cradle of Thought*.

12. Ibid.; Hobson, R. P., and A. Lee, "Imitation and Identification in Autism," *Journal of Child Psychology and Psychiatry* 40 (1999):649–59.

13. Williams, J. H., A. Whiten, T. Suddendorf, and D. I. Perrett, "Imitation, Mirror Neurons and Autism," *Neuroscience and Biobehavioral Review* 25 (2001):287–95.

14. Altschuler, E. L., A. Vankov, E. M. Hubbard, et al., "Mu Wave Blocking by Observation of Movement and Its Possible Use to Study the Theory of Other Minds," *Society for Neuroscience* (2000). Abstracts 68.1.

15. Nishitani, N., S. Avikainen, and R. Hari, "Abnormal Imitation-Related Cortical Activation Sequences in Asperger's Syndrome," *Annals of Neurology* 55 (2004):558–62.

16. Hari's paper (Nishitani, et al., "Abnormal Imitation-Related Cortical Activation Sequences in Asperger's Syndrome") was published in 2004. By

this time, Ramachandran and his colleagues had completed the mu wave suppression study for which their preliminary results had been presented at the big neuroscience meeting in 2000. This work also strongly suggests that individuals with autism are hindered by mirror neurons that are not fully functioning (Oberman, L. M., E. M. Hubbard, J. P. McCleery, et al., "EEG Evidence for Mirror Neuron Dysfunction in Autism Spectrum Disorders," *Brain Research: Cognitive Brain Research* 24 (2005):190–98). The Scottish group led by Justin Williams has also recently completed a study on adolescents with autism, using fMRI. When these adolescents imitate, the activity in mirror neuron areas is reduced compared with that of typically developing adolescents. This was the first evidence from brain imaging experiments supporting the hypothesis that the imitation deficits seen in patients with autism are indeed due to a reduced functioning of mirror neurons (Williams, J. H., G. D. Waiter, A. Gilchrist, et al., "Neural Mechanisms of Imitation and 'Mirror Neuron' Functioning in Autistic Spectrum Disorder," *Neuropsychologia* 44 (2006):610–21). Furthermore, Hugo Théoret's group in Montreal recently used TMS to test for a deficit in mirror neurons in subjects with autism. This experiment measured the excitability of the motor system while subjects watched the actions of other people. As detailed earlier, this excitability reflects a "motor resonance" mechanism that is considered another index of mirror neuron functioning, and Théoret and his colleagues found, not surprisingly, that individuals with autism had a much lower resonance than the healthy volunteers (Théoret, H., E. Halligan, M. Kobayashi, et al., "Impaired Motor Facilitation During Action Observation in Individuals with Autism Spectrum Disorder," *Current Biology* 15 (2005):R84–R85).

17. Dapretto, M., M. S. Davies, J. H. Pfeifer, et al., "Understanding Emotions in Others: Mirror Neuron Dysfunction in Children with Autism Spectrum Disorders," *Nature Neuroscience* 9 (2006):28–30.

18. Klin, A., W. Jones, R. Schultz, et al., "Visual Fixation Patterns During Viewing of Naturalistic Social Situations as Predictors of Social Competence in Individuals with Autism," *Archives of General Psychiatry* 59 (2002):809–16; Klin, A., W. Jones, R. Schultz, et al., "The Enactive Mind, or From Actions to Cognition: Lessons from Autism," *Philosophical Transactions of the Royal Society of London: B Biological Series* 358 (2003):345–60.

19. Field, T., C. Sanders, and J. Nadel, "Children with Autism Display More Social Behaviors After Repeated Imitation Sessions," *Autism* 5 (2001):317–23; Escalona, A., T. Field, J. Nadel, et al., "Brief Report: Imitation Effects on Children with Autism," *Journal of Autism and Developmental Disorders* 32 (2002):141–44.

20. Ingersoll, B., E. Lewis, and E. Kroman, "Teaching the Imitation and Spontaneous Use of Descriptive Gestures in Young Children with Autism Using a Naturalistic Behavioral Intervention," *Journal of Autism and Developmental Disorders* 37 (2007):1446–56; Ingersoll, B., and L. Schreibman, "Teaching Reciprocal Imitation Skills to Young Children with Autism Using a Naturalistic Behavioral Approach: Effects on Language, Pretend Play, and Joint Attention," *Journal of Autism and Developmental Disorders* 36 (2006):487–505; Ingersoll, B., and S. Gergans, "The Effect of a Parent-Implemented Imitation Intervention on Spontaneous Imitation Skills in Young Children with Autism," *Research and Developmental Disability* 28 (2007):163–75.

Seven: Super Mirrors and the Wired Brain

1. A good example is the collaborative paper between the lab of Giacomo Rizzolatti and the lab of Guy Orban, in which a monkey's brain activity during action observation is measured with fMRI rather than single-unit recordings: Nelissen, K., G. Luppino, W. Vanduffel, et al., "Observing Others: Multiple Action Representation in the Frontal Lobe," *Science* 310 (2005):332–36.

2. Leao, A.A.P., "Spreading Depression of Activity in the Cerebral Cortex," *Journal of Neurophysiology* 7 (1944):359–90; Leao, A.A.P., and R. S. Morrison, "Propagation of Spreading Cortical Depression," *Journal of Neurophysiology* 8 (1945):33–45.

3. Woods, R. P., M. Iacoboni, and J. C. Mazziotta, "Brief Report: Bilateral Spreading Cerebral Hypoperfusion During Spontaneous Migraine Headache," *New England Journal of Medicine* 331 (1994):1689–92.

4. Mukamel, R., H. Gelbard, A. Arieli, et al., "Coupling Between Neuronal Firing, Field Potential, and fMRI in Human Auditory Cortex," *Science* 309 (2005):951–54.

5. Gross, C. G., "Genealogy of the Grandmother Cell," *Neuroscientist* 8 (2002):512–18.

6. Gallese, V., L. Fadiga, L. Fogassi, et al., "Action Recognition in the Premotor Cortex," *Brain* 119 (Pt 2)(1996):593–609.

7. Quiroga, R. Q., L. Reddy, G. Kreiman, et al., "Invariant Visual Representation by Single Neurons in the Human Brain," *Nature* 435 (2005): 1102–07.

8. One possibility here is that the Jennifer Aniston cell codes the character of Rachel in *Friends*, rather than the actress Jennifer Aniston. This might explain why this cell does not fire at a picture of Aniston and Brad Pitt. Thanks to Kelsey Laird for suggesting this hypothesis.

9. Ekstrom, A. D., M. J. Kahana, J. B. Caplan, et al., "Cellular Networks Underlying Human Spatial Navigation," *Nature* 425 (2003):184–88; Kreiman, G., C. Koch, and I. Fried, "Imagery Neurons in the Human Brain," *Nature* 408 (2000):357–61.

10. Dijksterhuis, A., "Why We Are Social Animals: The High Road to Imitation as Social Glue," in Hurley and Chater, *Perspectives on Imitation, Volume 2*, 207–20.

11. Mukamel, R., A. D. Ekstrom, J. Kaplan, et al., "Mirror Neurons of Single Cells in Human Medial Frontal Cortex," Program No. 127.4 *2007 Abstract Viewer*, CD-ROM, Society for Neuroscience meeting, San Diego, CA.

Eight: The Bad and the Ugly: Violence and Drug Abuse

1. The first two sections of this chapter are based on my response to the 2006 *Edge*'s World Question (www.edge.org), also reprinted in J. Brockman, *What Is Your Dangerous Idea?: Today's Leading Thinkers on the Unthinkable* (London: Simon & Schuster, 2006), 71–74.

2. Brison, S., "Imitating Violence," in Hurley and Chater, *Perspectives on Imitation, Volume 2*, 202–204; Eldridge, J., "What Effects Does the Treatment of Violence in the Mass Media Have on People's Conduct? A Controversy Reconsidered," in Hurley and Chater, *Perspectives on Imitation, Volume 2*, 243–55.

3. Bandura, A., *Social Learning Theory* (Englewood Cliffs, NJ: Prentice Hall, 1977); Geen, R., and S. Thomas, "The Immediate Effects of Media Violence on Behaviour," *Journal of Social Issues* 42 (1986):7–28; Paik, H., and G. Comstock, "The Effects of Television Violence on Antisocial Behavior:

A Meta-analysis," *Communication Research* 21 (1994):516–46; Bushman, B., and L. Huesmann, "Effects of Television Violence on Aggression," in D. Singer and J. Singer, eds., *Handbook of Children and the Media* (Thousand Oaks, CA: Sage, 2001), 223–54.

4. Kostinsky, S., E. O. Bixler, and P. A. Kettl, "Threats of School Violence in Pennsylvania After Media Coverage of the Columbine High School Massacre: Examining the Role of Imitation," *Archives of Pediatric and Adolescent Medicine* 155 (2001):994–1001; Huesmann, L., and L. Eron, "Television and the Aggressive Child: A Cross-national Comparison," (Hillsdale, NJ: Erlbaum, 1986); Milavsky, J., R. Kessler, H. Stipp, et al., *Television and Aggression: A Panel Study* (New York: Academic Press, 1982).

5. Huesmann, L. R., "Imitation and the Effects of Observing Media Violence on Behavior," in Hurley and Chater, *Perspectives on Imitation, Volume 2*, 257–66.

6. Comstock, G., "Media Violence and Aggression, Properly Considered," in Hurley and Chater, *Perspectives on Imitation, Volume 2*, 371–80.

7. Hurley, S., "Imitation, Media Violence, and Freedom of Speech," *Philosophical Studies* 117 (2004):165–218; Brison, S., "Imitating Violence," in Hurley and Chater, *Perspectives on Imitation, Volume 2*, 202–204.

8. Marcus, S., *Neuroethics: Mapping the Field* (New York: Dana Press, 2002); Gazzaniga, M. S., *The Ethical Brain* (New York: Dana Press, 2005).

9. Maisto, S. A., and G. J. Connors, "Relapse in the Addictive Behaviors: Integration and Future Directions," *Clinical Psychology Review* 26 (2006):229–31; Gordon, S. M., R. Sterling, C. Siatkowski, et al., "Inpatient Desire to Drink as a Predictor of Relapse to Alcohol Use Following Treatment," *American Journal of Addiction* 15 (2006):242–45; Shiffman, S., J. A. Paty, M. Gnys, et al., "First Lapses to Smoking: Within-Subjects Analysis of Real-time Reports," *Journal of Consulting in Clinical Psychology* 64 (1996):366–79; Harakeh, Z., R. C. Engels, R. B. Van Baaren, et al., "Imitation of Cigarette Smoking: An Experimental Study on Smoking in a Naturalistic Setting," *Drug and Alcohol Dependence* 86 (2007):199–206.

10. Calvo-Merino, B., D. E. Glaser, J. Grèzes, et al., "Action Observation and Acquired Motor Skills: An fMRI Study with Expert Dancers," *Cerebral*

Cortex 15 (2005):1243–49; Calvo-Merino, B., J. Grèzes, D.E. Glaser et al., "Seeing or Doing? Influence of Visual and Motor Familiarity in Action Observation," *Current Biology* 16 (2006):1905–10; Shiraishi, T., H. Saito, H. Ito, et al., "Observation and Imitation of Nursing Actions: A NIRS Study with Experts and Novices," *Student Health and Technology Information* 122 (2006):820–21.

Nine: Mirroring Wanting and Liking

1. A good review of these studies is in Schooler, J. W., "Re-presenting Consciousness: Dissociations Between Experience and Meta-consciousness," *Trends in Cognitive Sciences* 6 (2002):339–44.
2. Johansson, P., L. Hall, S. Sikstrom, et al., "Failure to Detect Mismatches Between Intention and Outcome in a Simple Decision Task," *Science* 310 (2005):116–19.
3. Schultz, W., P. Dayan, and P. R. Montague, "A Neural Substrate of Prediction and Reward," *Science* 275 (1997):1593–99; Montague, P. R., B. King-Casas, and J. D. Cohen, "Imaging Valuation Models in Human Choice," *Annual Review of Neuroscience* 29 (2006):417–48.
4. McClure, S. M., J. Li, D. Tomlin, et al., "Neural Correlates of Behavioral Preference for Culturally Familiar Drinks," *Neuron* 44 (2004):379–87.

Ten: Neuropolitics

1. Converse, P., "The Nature of Belief Systems in Mass Publics," in D. Apter, ed., *Ideology and Discontent* (New York: Free Press, 1964), 206–61; Achen, C., "Mass Political Attitudes and the Survey Response," *American Political Science Review* 69 (1975):1218–31; Zaller, J. R., and S. Feldman, "A Simple Theory of the Survey Response: Answering Questions versus Revealing Preferences," *American Journal of Political Science* 36 (1992):579–616.
2. Raichle, M. E., J. A. Fiez, T. O. Videen, et al., "Practice-Related Changes in Human Brain Functional Anatomy During Nonmotor Learning," *Cerebral Cortex* 4 (1994):8–26.
3. Carr, L., M. Iacoboni, M. C. Dubeau, et al., "Neural Mechanisms of Empathy in Humans: A Relay from Neural Systems for Imitation to Limbic Areas," *Proceedings of the National Academy of Sciences USA* 100 (2003):5497–5502.

4. Schreiber, D., and M. Iacoboni, "Monkey See, Monkey Do: Mirror Neurons, Functional Brain Imaging, and Looking at Political Faces," paper presented at the American Political Science Association Meeting, 2005, Washington, D.C.

5. Gusnard, D. A., and M. E. Raichle, "Searching for a Baseline: Functional Imaging and the Resting Human Brain," *Nature Reviews Neuroscience* 2 (2001):685–94; Raichle, M. E., A. M. MacLeod, A. Z. Snyder, et al., "A Default Mode of Brain Function," *Proceedings of the National Academy of Sciences USA* 98 (2001):676–82.

6. Schreiber, D., and M. Iacoboni, "Thinking About Politics: Results from Three Experiments Studying Sophistication," paper presented at the 61st Annual National Conference of the Midwest Political Science Association, 2003.

7. Iacoboni, M., M. D. Lieberman, B. J. Knowlton, et al., "Watching Social Interactions Produces Dorsomedial Prefrontal and Medial Parietal BOLD fMRI Signal Increases Compared to a Resting Baseline," *Neuroimage* 21 (2004):1167–73.

8. Fiske, A. P., *Structures of Social Life: The Four Elementary Forms of Human Relations* (New York: Free Press, 1991).

9. Iacoboni, M., "Failure to Deactivate in Autism: The Co-constitution of Self and Other," *Trends in Cognitive Science* 10 (2006):431–33; Uddin, L. Q., M. Iacoboni, C. Lange, and J. P. Keenan, "The Self and Social Cognition: The Role of Cortical Midline Structures and Mirror Neurons," *Trends in Cognitive Science* 11 (2007):153–57; Lieberman, M. D., "Social Cognitive Neuroscience: A Review of Core Processes," *Annual Review of Psychology* 58 (2007):259–89.

Eleven: Existential Neuroscience and Society

1. One day I told this story to Giacomo Rizzolatti. He said that he had read similar remarks in a newspaper interview with Peter Brook, the world-renowned theater director. Is this story another meme with high replicability?

2. Wittgenstein, L., *Remarks on the Philosophy of Psychology*, Volume 2 (Oxford, UK: Blackwell, 1980); Merleau-Ponty, M., *The Primacy of Perception* (Evanston, IL: Northwestern University Press, 1964).

3. Benner, P., "The Quest for Control and the Possibilities of Care," in M. Wrathall and J. Malpas, eds., *Heidegger, Coping, and Cognitive Science: Essays in Honor of Hubert L. Dreyfus, Volume 2* (Cambridge, MA: MIT Press, 2000), 293–309.

4. Heidegger, M., *Being and Time* (New York: Harper & Row, 1962); Sartre, J.-P., *Being and Nothingness: A Phenomenological Essay on Ontology* (New York: Citadel Press, 1956).

5. Indeed, the philosopher Hubert Dreyfus, in his presidential address to the Pacific Division of the American Philosophical Association, emphasized what is wrong with the analytic/continental dichotomy and forcefully reminded us why both "sides" of philosophy are important. Dreyfus, H. L., "Overcoming the Myth of the Mental: How Philosophers Can Profit from the Phenomenology of Everyday Expertise," APA Pacific Division Presidential Address, 2005.

6. This final section of the book is partly based on my response to the 2007 *Edge* World Question (www.edge.org) "What are you optimistic about? Why?"

7. Heidegger, M., *Being and Time*; Zahavi, D., "Beyond Empathy," *Journal of Consciousness Studies* 8 (2001):151–67.

8. Olson, G., "Hard-wired for Moral Politics: Neuroscience and Empathy," *ZNet* (www.zmag.org), May 20, 2007; Amin, A., "From Ethnicity to Empathy: A New Idea of Europe," *openDemocracy* (www.opendemocracy.net), July 23, 2003; Olson, G., "Neuroscience and moral politics: Chomsky's intellectual progeny," Identitytheory.com (www.identitytheory.com/social/olson_neuro.php), October 16, 2007.

Acknowledgments

This book would not have been possible without the help, encouragement, and support of many friends and colleagues. First and foremost I thank John Brockman for his unfaltering encouragement to write it. I also thank Katinka Matson, Mike Bryan, and my editor, Eric Chinski, for shaping the manuscript in many important ways.

Several people read individual chapters from early and late drafts of the book. I am grateful to George Lakoff, Sam Harris, Annaka Harris, Frank Vincenzi, Sally Rogers, Kelsey Laird, Amy Coplan, Lisa Aziz-Zadeh, Elizabeth Reynolds, Julian Keenan, Alan Fiske, John Mazziotta, Giacomo Rizzolatti, and Vittorio Gallese for their comments, suggestions, and questions.

A common thread of the book is the research that has been performed in my lab in the last ten years. This research has been possible because of the dedication and enthusiasm of colleagues and trainees. First and foremost, I am in debt to Giacomo Rizzolatti and Vittorio Gallese, wonderful friends and colleagues who participated in seminal experiments performed in my lab. John Mazziotta, Roger Woods, Harold Bekkering, Marcel Brass, Andreas Wohlschläger, Eran Zaidel, Gian Luigi Lenzi, Patricia Greenfield, and Itzhak Fried also participated in crucial experiments on the human mirror neuron system. With her own lab, my wife and colleague, Mirella Dapretto, led groundbreaking research on mirror neuron dysfunction in autism, and I was fortunate to collaborate on these studies.

I celebrate my trainees, who have enriched my life in many ways. Mentoring and doing experiments with them has been illuminating and exhilarating: Lisa Aziz-Zadeh, Laurie Carr, Choi Deblieck, Marie-Charlotte Dubeau, Marc Heiser, Jonas Kaplan, Lisa Koski, Ingo Meister, Istvan Molnar-Szakacs, Roy Mukamel, Darren Schreiber, Lucina Uddin, Stephen Wilson, and Allan Wu were involved in experiments and endless discussions on how mirror neurons shape our social behavior.

With leadership and vision, John Mazziotta created a wonderful research facility—the Ahmanson-Lovelace Brain Mapping Center—where my lab is located. I celebrate John and the center and feel fortunate to do my research in such a world-class facility. I thank the UCLA Semel Institute for Neuroscience and Human Behavior and the FPR-UCLA Center for Culture, Brain, and Development for creating extraordinarily stimulating environments where I often discussed the role of mirror neurons in human behavior.

In the last ten years, I have given seminars on mirror neurons all over the world. I thank all the people who came to hear me speak, asked me questions, and gave me comments. All those people helped in shaping the arguments I put in this book. I am deeply grateful to them.

Index

Page references in *italics* refer to illustrations.